Smart-Kids

Gené Peters, Carol Every

Mathematics

英汉对照

奔跑吧数学

4级
Grade 4

探索奇妙的数学世界

〔英〕 吉恩·彼得斯
卡罗尔·艾瑞 主编
许译丹 译

天津出版传媒集团
天津科技翻译出版有限公司

著作权合同登记号:图字:02-2016-13

图书在版编目(CIP)数据

奔跑吧,数学:探索奇妙的数学世界.4级:英汉对照/(英)吉恩·彼得斯(Gené Peters),(英)卡罗尔·艾瑞(Carol Every)主编;许译丹译.—天津:天津科技翻译出版有限公司,2016.8

书名原文:Smart-Kids Mathematics:Grade 3

ISBN 978-7-5433-3627-8

Ⅰ.①奔… Ⅱ.①吉… ②卡… ③许… Ⅲ.①数学-儿童读物-英、汉 Ⅳ.①O1-49

中国版本图书馆CIP数据核字(2016)第168307号

出　　版:天津科技翻译出版有限公司
出 版 人:刘 庆
地　　址:天津市南开区白堤路244号
邮政编码:300192
电　　话:(022)87894896
传　　真:(022)87895650
网　　址:www.tsttpc.com
印　　刷:天津市银博印刷集团有限公司
发　　行:全国新华书店
版本记录:880×1230　16开本　6.25印张　100千字
2016年8月第1版　2016年8月第1次印刷
定价:29.80元

(如发现印装问题,可与出版社调换)

出版者的话

《奔跑吧，数学：探索奇妙的数学世界》(英汉双语)(1~4级)是从国际著名教育出版机构英国培生教育集团引进的数学学习益智书，真实反映了国外小学生的现行教学内容，全面展现了国外小学生丰富多彩的学习场景。

为什么很多小学生不喜欢学习数学，学习效果不好，没有学习的兴趣？这恐怕和我们侧重于背公式，做习题，准备考试，这种比较枯燥的学习方式不无关系。这套丛书全面体现国外小学生要掌握的数学基础知识和英语表达，展现生动活泼的学习和游戏场景。读者可从中领会原汁原味的国外小学生的学习内容，学习简单的英语表达。同时，书中着重通过游戏让孩子亲自动手，寓教于乐，图文并茂，让孩子在提高动手能力的同时提高学习数学的兴趣。通过游戏的方式，让孩子在奇妙的数学世界中快乐地奔跑、遨游和探索。

书后备有小贴纸，可以增加孩子学习的乐趣。并贴心地配有"注释"，介绍了每个小游戏的训练目的和训练方法，帮助孩子和家长一起打开学习数学的大门。每册学习完成后，家长可以为孩子颁发"证书"，让孩子拥有满满的成就感。

这套丛书采用英汉双语对照的形式，既保留了原版英文，介绍了原汁原味的英语背景和地道的英语口语表达，又为方便孩子理解可以独立完成练习而增加了中文翻译，在学习数学技能、提高数学学习能力的同时，也能提高英语水平，可谓一书在手，一举两得。

目录

CONTENTS

Term 1

Term 2

Term 3

Term 4

奔跑吧，数学

英汉对照
4级

探索奇妙的数学世界

Smart-Kids

Mathematics Grade 4

Follow the paw prints 跟着脚印走

The animals went for a walk in the forest. 动物们在森林里散步。

1. Write the numbers in each path of paw prints. 写出每条路线的脚印数。

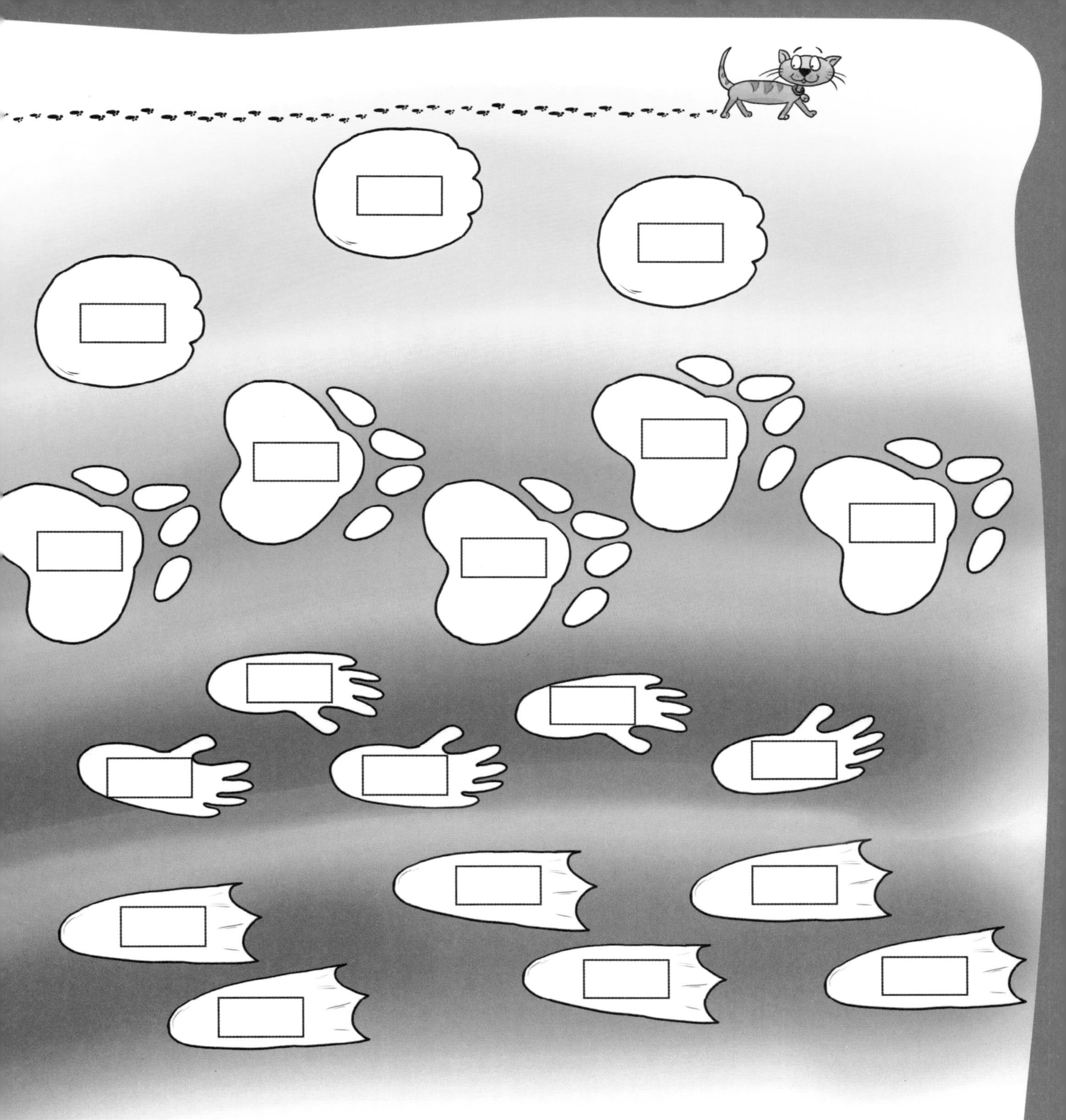

2. Colour the paw prints that contain the answers to these problems.
将包含下列习题答案的那些脚印涂色。

$10 \times 4 =$	$2 \times 5 =$	$2 \times 3 =$
$3 \times 6 =$	$3 \times 5 =$	$2 \times 10 =$
$3 \times 3 =$	$10 \times 1 =$	$5 \times 3 =$
$10 \times 6 =$	$5 \times 5 =$	$5 \times 1 =$

Lebo’s Share-a-lot

Lebo 均分水果

Lebo’s mom asked her to share some fruit with her friends. Can you help her?

Lebo 的妈妈让她和朋友们一起分享一些水果。
你可以帮助她吗？

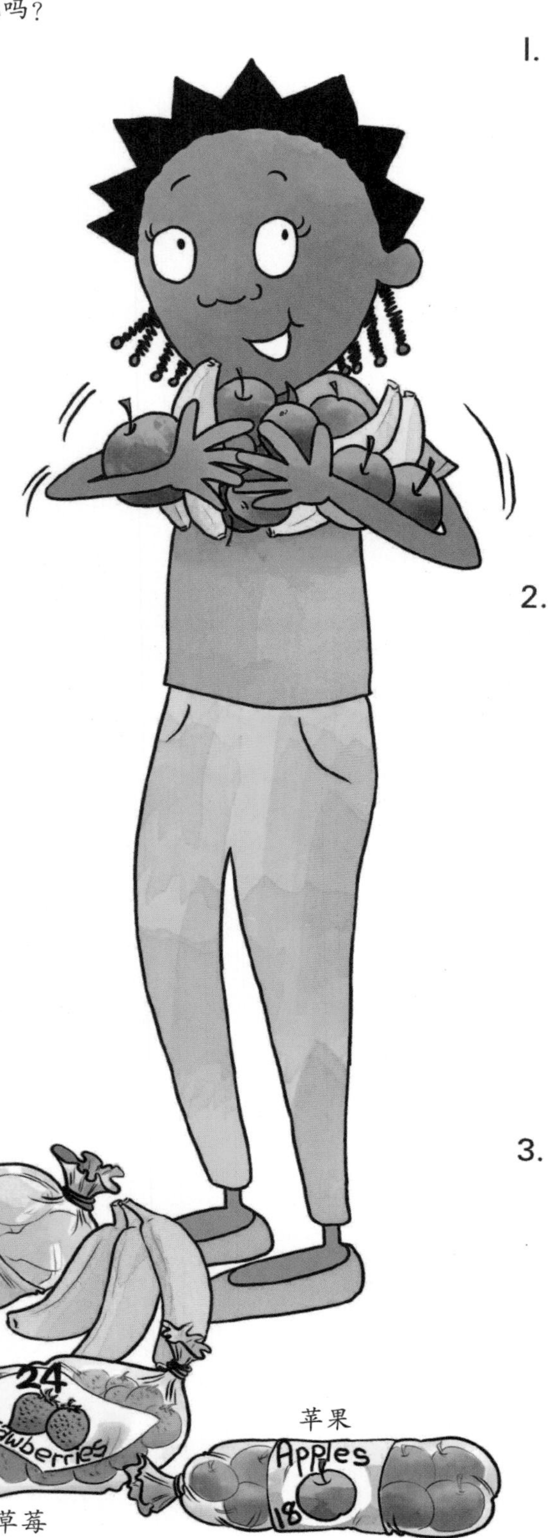

1. 20 ÷ 5 = ☐
   20 ÷ 2 = ☐
   20 ÷ 4 = ☐

2. 18 ÷ 3 = ☐
   18 ÷ 6 = ☐
   18 ÷ 9 = ☐

3. 24 ÷ 2 = ☐
   24 ÷ 6 = ☐
   24 ÷ 4 = ☐

See-saw sums
跷跷板算术题

Can you help the seal balance the balls?
你可以帮助海豹保持这些球的平衡吗？

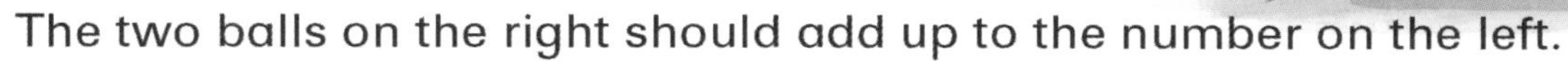

The two balls on the right should add up to the number on the left.
右边的 2 个球上面的数字相加得到的数应与左边的球上面的数字相同。

Skill: subtraction

Collecting money

募捐

The children are collecting money for sick animals.
Can you work out who has collected the most money?

孩子们为生病的动物募捐。你可以算出谁捐得最多吗？

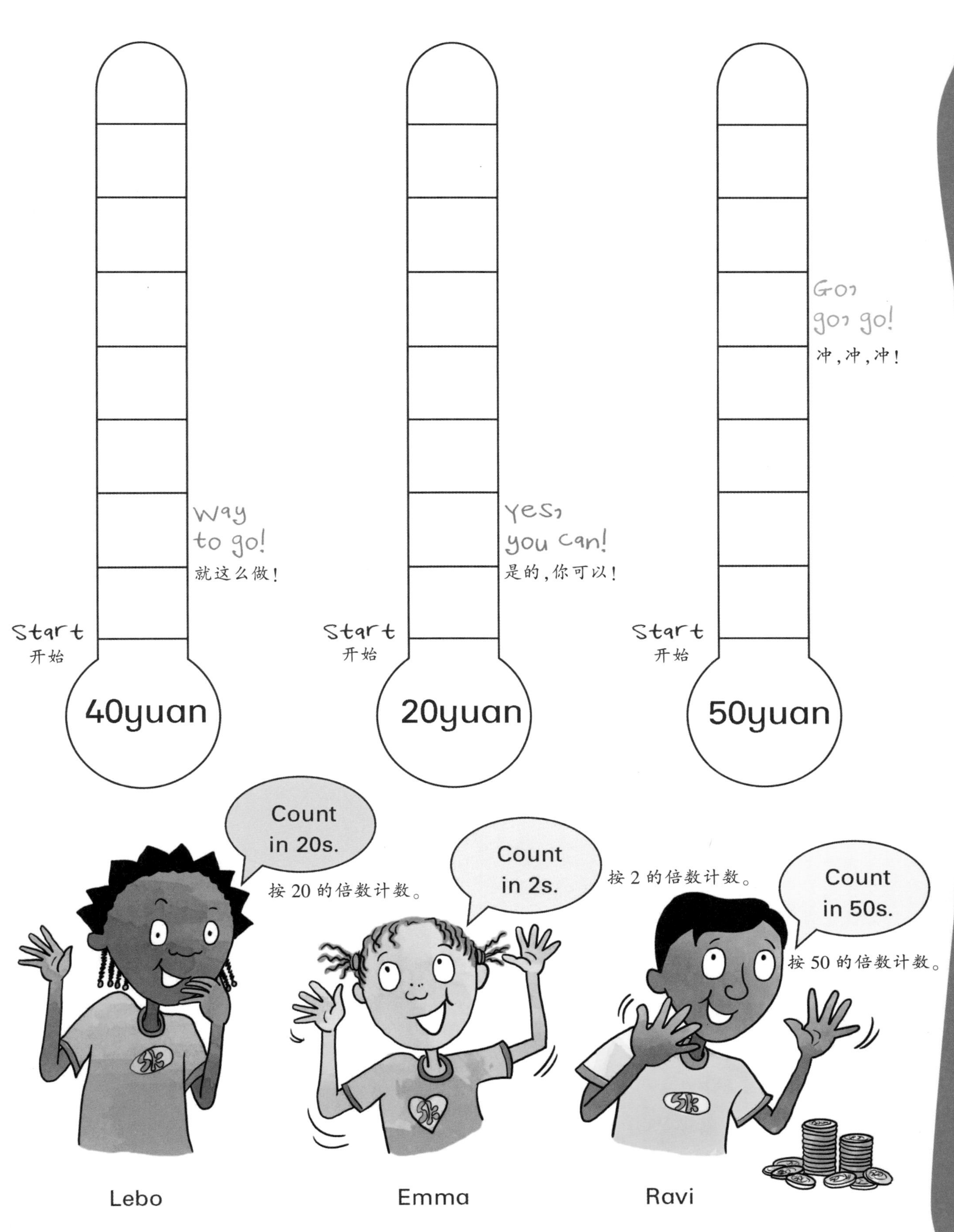

Who collected the most money ? ____________________

谁募捐得钱最多？

Skill: counting in multiples

Quiz master
智力竞赛节目主持人

1. Complete each quiz.
   完成测验。

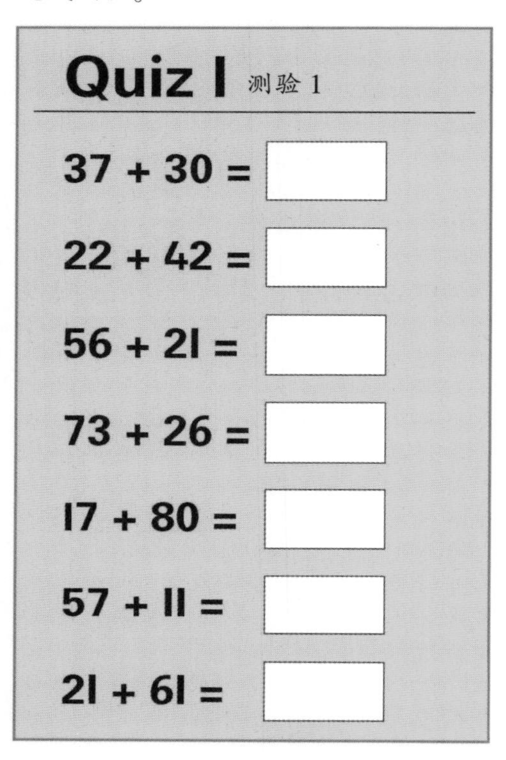

Quiz 1 测验 1

37 + 30 = ☐

22 + 42 = ☐

56 + 21 = ☐

73 + 26 = ☐

17 + 80 = ☐

57 + 11 = ☐

21 + 61 = ☐

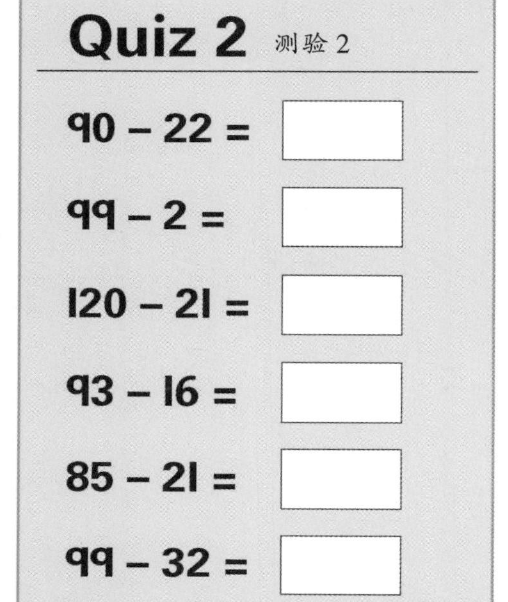

Quiz 2 测验 2

90 – 22 = ☐

99 – 2 = ☐

120 – 21 = ☐

93 – 16 = ☐

85 – 21 = ☐

99 – 32 = ☐

99 – 17 = ☐

2. Draw a line between the sums that have the same answer.
   将答案相同的算术题连线。
3. Colour the answers in the grid below.
   在下面方框中，将含有上述答案的方框涂色。

16	12	91	32	46
96	82	68	64	51
85	18	97	73	89
21	67	99	77	24
45	91	33	71	61

4. What letter do you see? ____________________
   你看到了哪个字母？

Pirate loot
海盗战利品

Help Pirate Pete to divide the loot between himself and the other six men on board.
帮助海盗 Pete 和船上的其他 6 名海盗分配战利品。

1. There are 40 gold coins. Pete first takes $\frac{1}{2}$ of the coins for himself.

   有 40 枚金币。Pete 首先为自己拿了金币的$\frac{1}{2}$。
   a. How many coins does Pete get? ☐
   Pete 获得了多少金币?
   b. How many coins are left? ☐
   剩下多少金币?

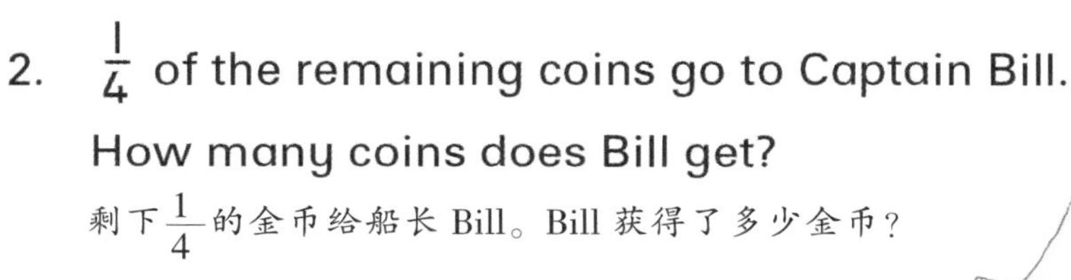

2. $\frac{1}{4}$ of the remaining coins go to Captain Bill. How many coins does Bill get?
   剩下$\frac{1}{4}$的金币给船长 Bill。Bill 获得了多少金币?

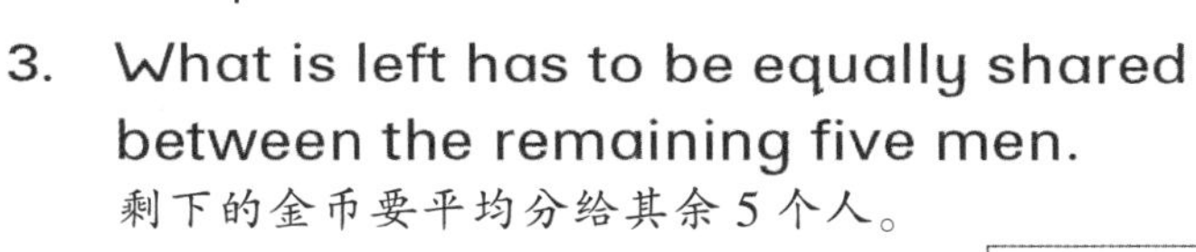

3. What is left has to be equally shared between the remaining five men.
   剩下的金币要平均分给其余 5 个人。
   a. How many coins are left? ☐
   剩下多少金币?
   b. How many coins will each get? ☐
   每个人获得多少金币?

4. Who got the most? Who got the fewest? From the biggest to the smallest, write down the number names for the number of coins Pete, Bill and each of the remaining pirates get.
   谁获得的金币最多?谁获得的金币最少?按照从大到小的顺序写出 Pete、Bill 和其余每个海盗获得的金币数。

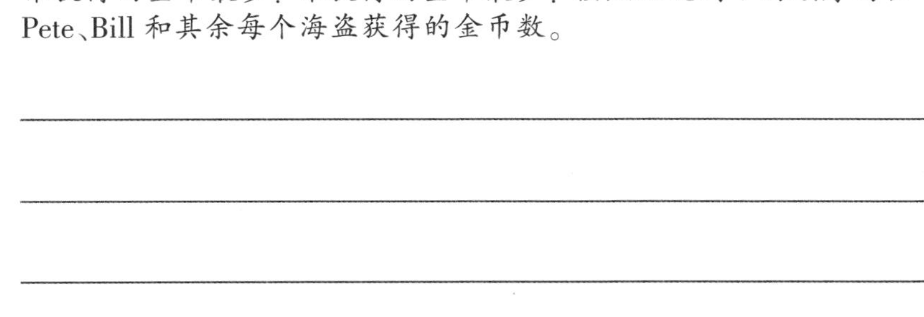

____

____

____

5. Circle the correct fraction. 圈出正确的分数。
   Bill received $\frac{1}{2}$ / $\frac{1}{3}$ / $\frac{1}{4}$ of the coins Pete took.
   Bill 拿到的金币数是 Pete 的$\frac{1}{2}$/$\frac{1}{3}$/$\frac{1}{4}$。

Term 1

Skill: working with fractions

Mini-beast mania

昆虫迷

Jaco loves collecting mini-beasts. Here are the ones in his collection. Jaco 喜欢收集昆虫。这是他的一些收藏。

1. Count how many of each you can see in the pictures above.
数一数,你在上面图片中看到下列每个部位各有多少。

mini-beasts	legs	feelers	spots	wings
昆虫	腿	触角	斑点	翅膀

2. Complete the graph by sticking your stickers in the blocks to show how many you counted. 用贴纸在表格上展示出你数出多少,完成图表。

14					
13					
12					
11					
10					
9					
8					
7					
6					
5					
4					
3					
2					
1					
	mini-beasts 昆虫	legs 腿	feelers 触角	spots 斑点	wings 翅膀

Time for time 时间

Are you learning to tell the time?

你学会描述时间了吗？

1. Say the time on each clock.
   说出每个钟表上的时间。
2. Write it down.
   写出时间。

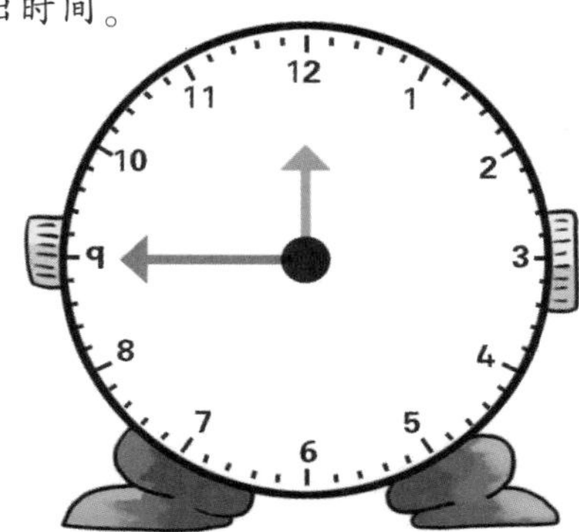

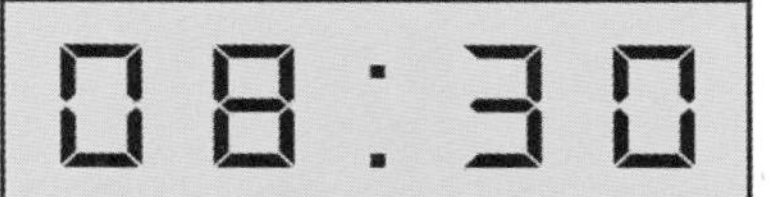

3. Draw the hands on the clock below to show the time.
   在下面钟表上画出指针来显示时间。

Ten to five
4:50

4. Write the time on this clock.
   写出这个钟表的时间。

Quarter past two in the afternoon
下午 2 点 15 分

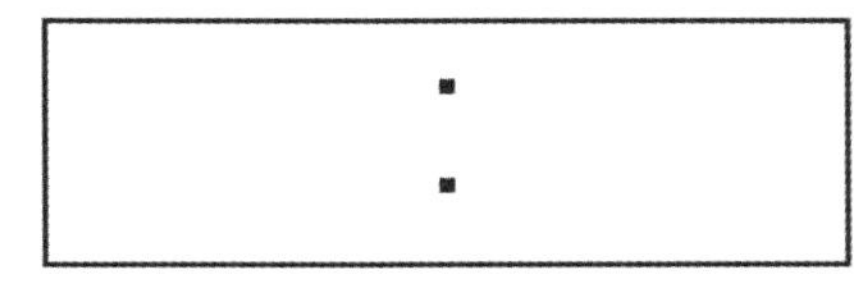

Shopping supplies

购物单

Jody, Mandla, Jaco and Ravi made a list when they decided to go fishing. Jody、Mandla、Jaco 和 Ravi 列了一个清单,准备去钓鱼。

1. Another four friends joined them. Rewrite the shopping list so that there will be enough supplies for all eight children. Remember to double everything on the list. 又有 4 个朋友加入他们。重新列出购物清单,以便有足够的用品给这 8 个孩子。记住清单上的每个物品都要加倍。

Shopping list 购物清单

4 fishing rods 4 个钓鱼竿

16 hooks 16 个挂钩

24 worms 24 只幼虫

8 eggs 8 个鸡蛋

18 rolls 18 个小面包

14 cooldrinks 14 瓶饮料

12 chocolates 12 块巧克力

New list 新的清单

_______ fishing rods 钓鱼竿

_______ hooks 挂钩

_______ worms 幼虫

_______ eggs 鸡蛋

_______ rolls 小面包

_______ cooldrinks 饮料

_______ chocolates 巧克力

2. The next time the children planned to go fishing, Mandla and Ravi were sick. Can you rewrite the list so that there are enough things for two children? Remember to halve everything on the list.
孩子们开始计划下次钓鱼,Mandla 和 Ravi 生病了。你可以重新为 2 个孩子列出新的清单吗?记住将清单上的每个物品减半。

New list 新的清单

_______ fishing rods 钓鱼竿

_______ hooks 挂钩

_______ worms 幼虫

_______ eggs 鸡蛋

_______ rolls 小面包

_______ cooldrinks 饮料

_______ chocolates 巧克力

Fill it up 装满

Measuring liquids can be fun.
Try not to spill!
测量液体很有趣。尽量不要溢出来！

Find these things for measuring:
找出这些东西用来测量：

1. Fill the basin with water. Use this water to measure with.
   将盆装满水。用这些水来测量。
2. Complete the table below. Remember to estimate before you measure. Choose your own containers for the last measurement.
   完成下面的表格。记住在测量前要估计一下。最后用你自己挑选的容器测量。

How many ... 多少……	Estimate 估计值	Measure 测量值
tablespoons of water fill the mug? 多少汤匙水能装满杯子？		
mugs of water fill the bowl? 多少杯水能装满碗？		
bowls of water fill the jug? 多少碗水能装满水罐？		
jugs of water fill the bucket? 多少罐水能装满水桶？		
buckets of water fill the basin? 多少桶水能装满盆？		

Pretty patterns 漂亮的图案

Finish the pretty patterns. 完成这些漂亮的图案。

1. Use your stickers to complete the patterns.
   用贴纸完成图案。
2. Which shapes have round sides? Colour their frames red.
   哪些图形有圆边？将它们的边框涂成红色。
3. Which shapes have straight sides? Colour their frames green.
   哪些图形有直线边？将它们的边框涂成绿色。

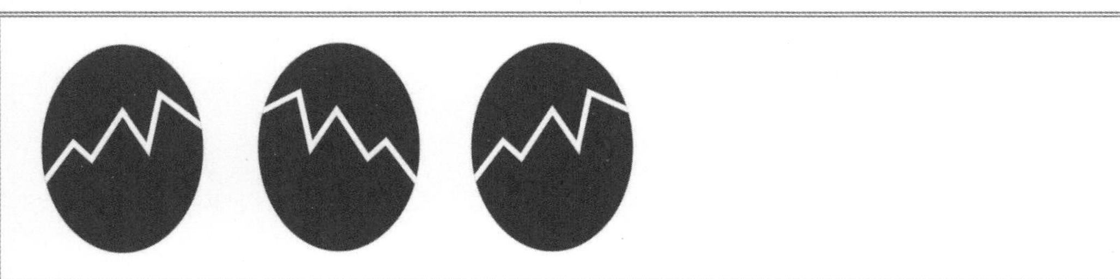

Picnic in the park 在公园野餐

Four friends are having a picnic.
4位伙伴正在野餐。

1. Count the picnic treats.
   数一数野餐的种类。
2. Now share the treats between four friends. You can halve the fruit juice to share it. How many will each friend get?
   现在这4位伙伴互相分享食物。你可以将果汁平分。每个朋友分了多少？
3. How many will be left over?
   剩下的是多少？

	Count 总计	Each share 每人分到	Left over 剩下的
hot dogs 热狗			
apples 苹果			
packets of chips 薯片			
juice 果汁			
cupcakes 蛋糕			
oranges 橙子			

Skill: sharing – division with remainders

Ready, steady, go

各就各位，预备——跑

Have you seen racing cars speed around a race track?
See how quickly you can complete each race track.
Ask an adult to time you.

你看到赛车在赛道上的速度了吗？看看你完成每条赛道的速度有多快。请一位大人为你计时。

1. Write the time in the stopwatch at the end of each race track.

   写出到达每条赛道终点时，秒表上的时间。

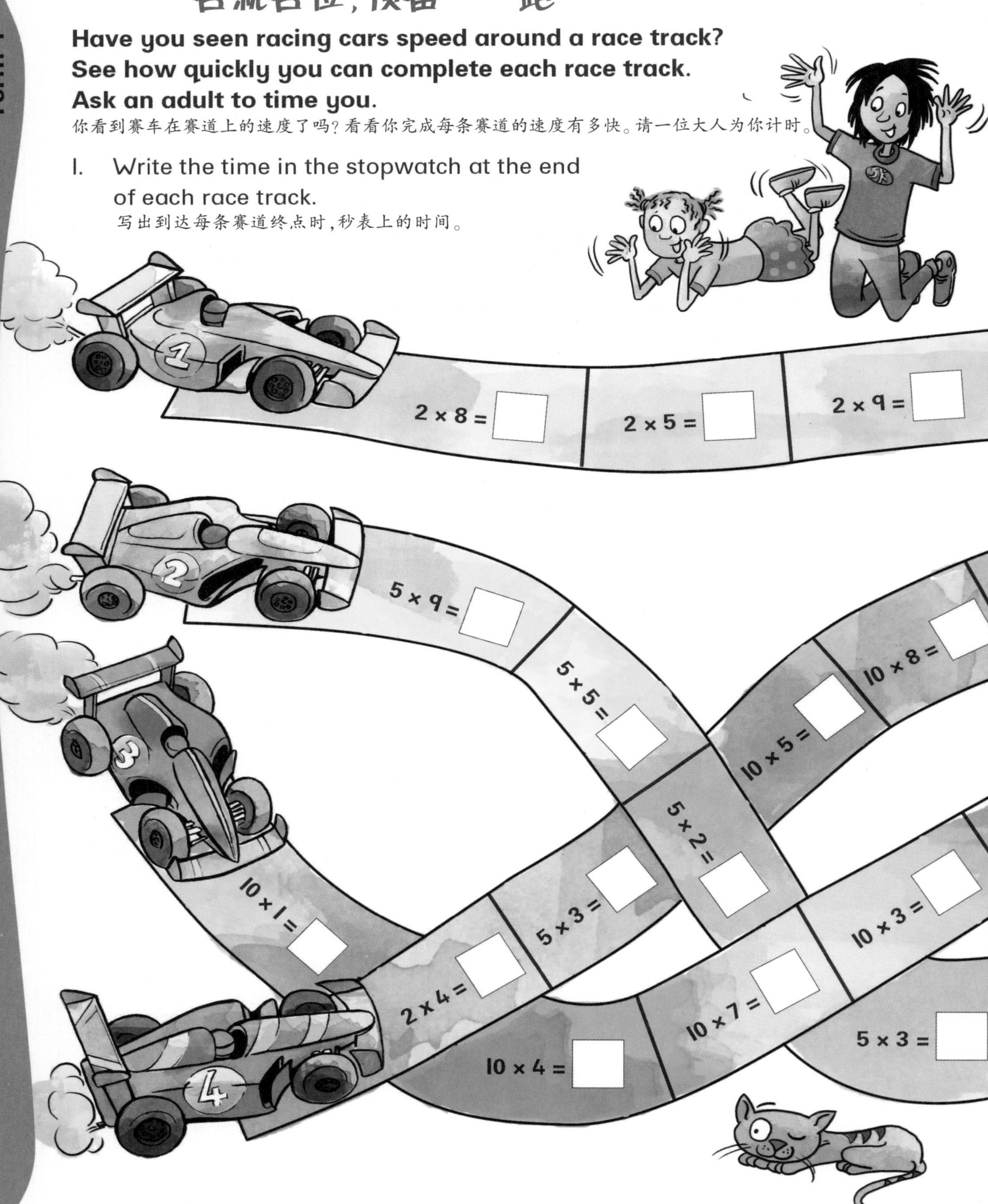

2. Which race was your fastest? Colour the stopwatch orange.
哪条赛道是你完成得最快的？将该秒表涂成橙色。
3. Which race was your slowest? Colour the stopwatch brown.
哪条赛道是你完成得最慢的？将该秒表涂成棕色。

Skill: multiplication

Shape hide-and-seek

图形捉迷藏

We hope this puzzle does not make you dizzy!

我们希望这个谜题不会让你头晕！

1. Find the shapes.
   找出各种图形。

2. Complete the sentences. 完成这些句子。

a. I can see ☐ triangles.
我能看见 ☐ 个三角形。

b. I can see ☐ squares.
我能看见 ☐ 个正方形。

c. I can see ☐ rectangles.
我能看见 ☐ 个长方形。

d. I can see ☐ circles.
我能看见 ☐ 个圆形。

e. These shapes have straight edges: ____________________
这些图形含有直线边：

f. These shapes have curved edges: ____________________
这些图形含有圆边：

g. These shapes have the same number of sides: ____________________
这些图形有相同的边数：

____

Posh pet 时髦的宠物

It takes time and costs money to care for your pet. Help Emma work out how much Stinky costs her family in one year.

照顾宠物需要花费时间和钱。帮助 Emma 计算出 Stinky 一年花了家里多少钱。

	How often 次数	Price 价格	Total per year 年总计
Dog shampoo 洗发水	Four times a year 1 年 4 次	10yuan	
Flea powder 虱子粉	Twice a year 1 年 2 次	17.5yuan	
Grooming parlour 美容院打理	Twice a year 1 年 2 次	90yuan	
Toys 玩具	Once a year 1 年 1 次	100yuan	
Dog biscuits 狗饼干	Once a month 1 个月 1 次	25yuan	
Vet bills 兽医账单	Twice a year 1 年 2 次	100yuan	

1. Complete the table. Work out the total cost in one year.
   完成表格。计算出一年总共的花费。
2. If Emma wants to take the exact money to pay the dog parlour, what is the least number of notes she should take? Draw them.
   如果 Emma 想把钱花在狗的美容上，她最少要用几张纸币刚好够这个钱数？把它们画出来。

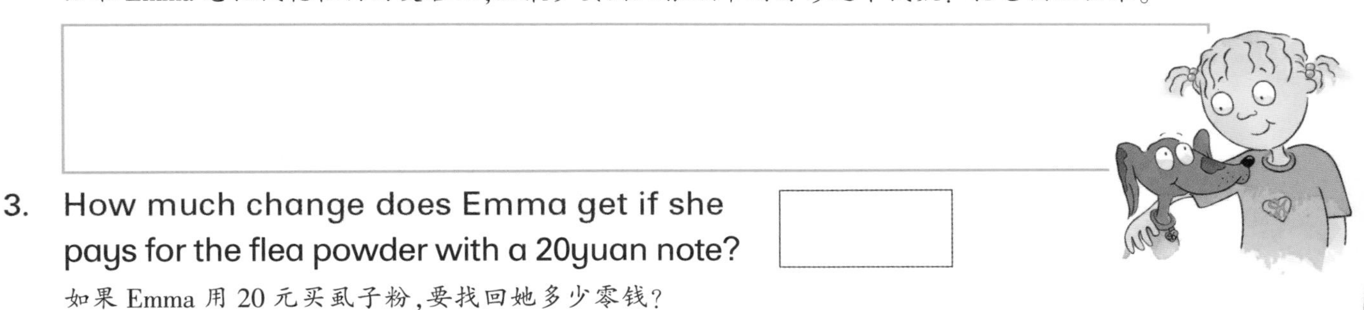

3. How much change does Emma get if she pays for the flea powder with a 20yuan note?
   如果 Emma 用 20 元买虱子粉，要找回她多少零钱？
4. What change will Emma get if she uses 30yuan to pay for dog biscuits? Draw the coins you think she will get back.
   如果 Emma 用 30 元买狗饼干，要找回她多少零钱？画出你认为应找回她的硬币。
5. What is the most expensive yearly cost in the table? ____________
   表格中一年花费最多的是？
6. What is the cheapest yearly cost in the table? ____________
   表格中一年花费最少的是？

Space race 太空竞赛

Race to your spaceship before the alien catches you with its tentacles!

在外星人用触角抓住你之前,要跑到你的飞船上!

Follow the instruction at the bottom of each pathway to reach the spaceship.

按照每条路线下方的指令到达飞船。

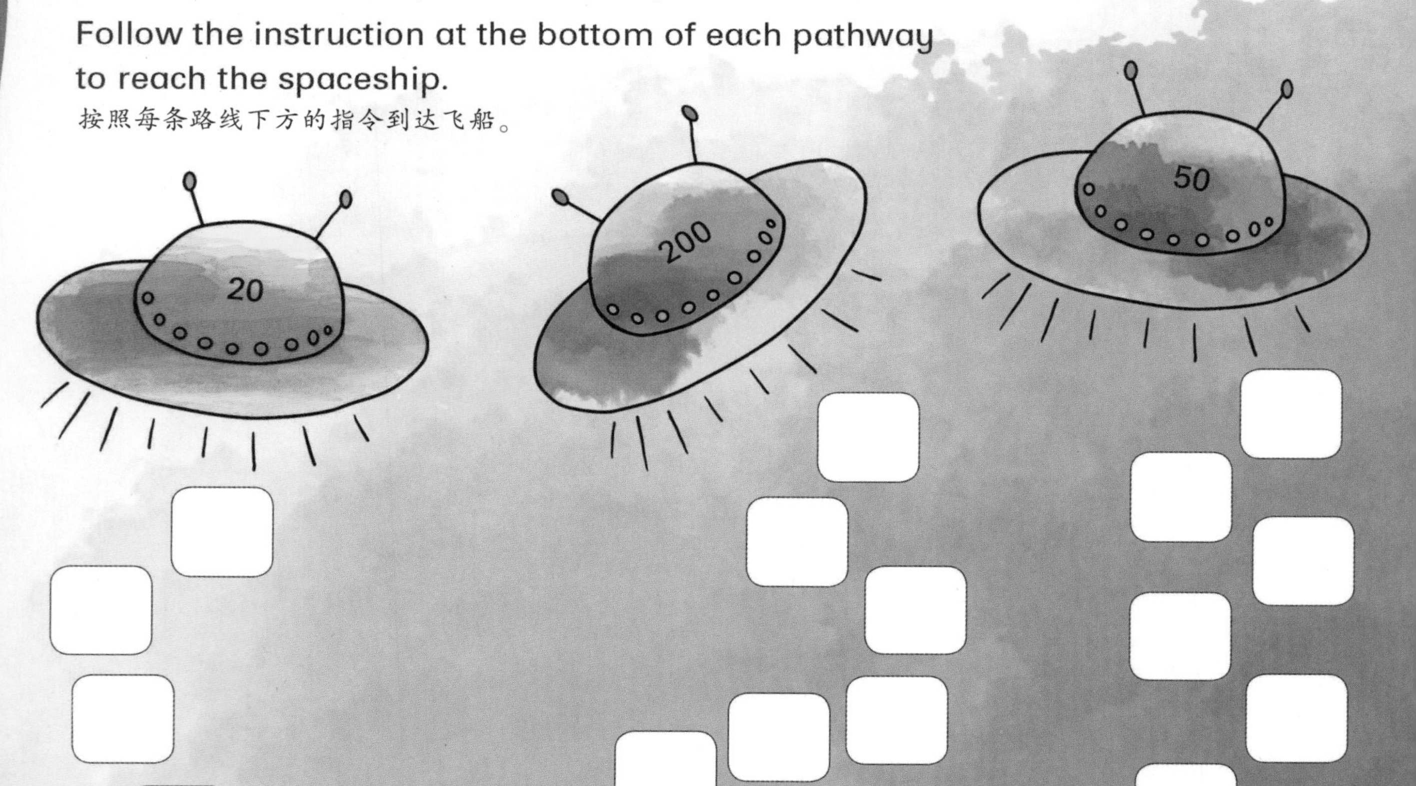

Count in 2s.

按 2 的倍数计数。

Count in 20s.

按 20 的倍数计数。

Count in 5s.

按 5 的倍数计数。

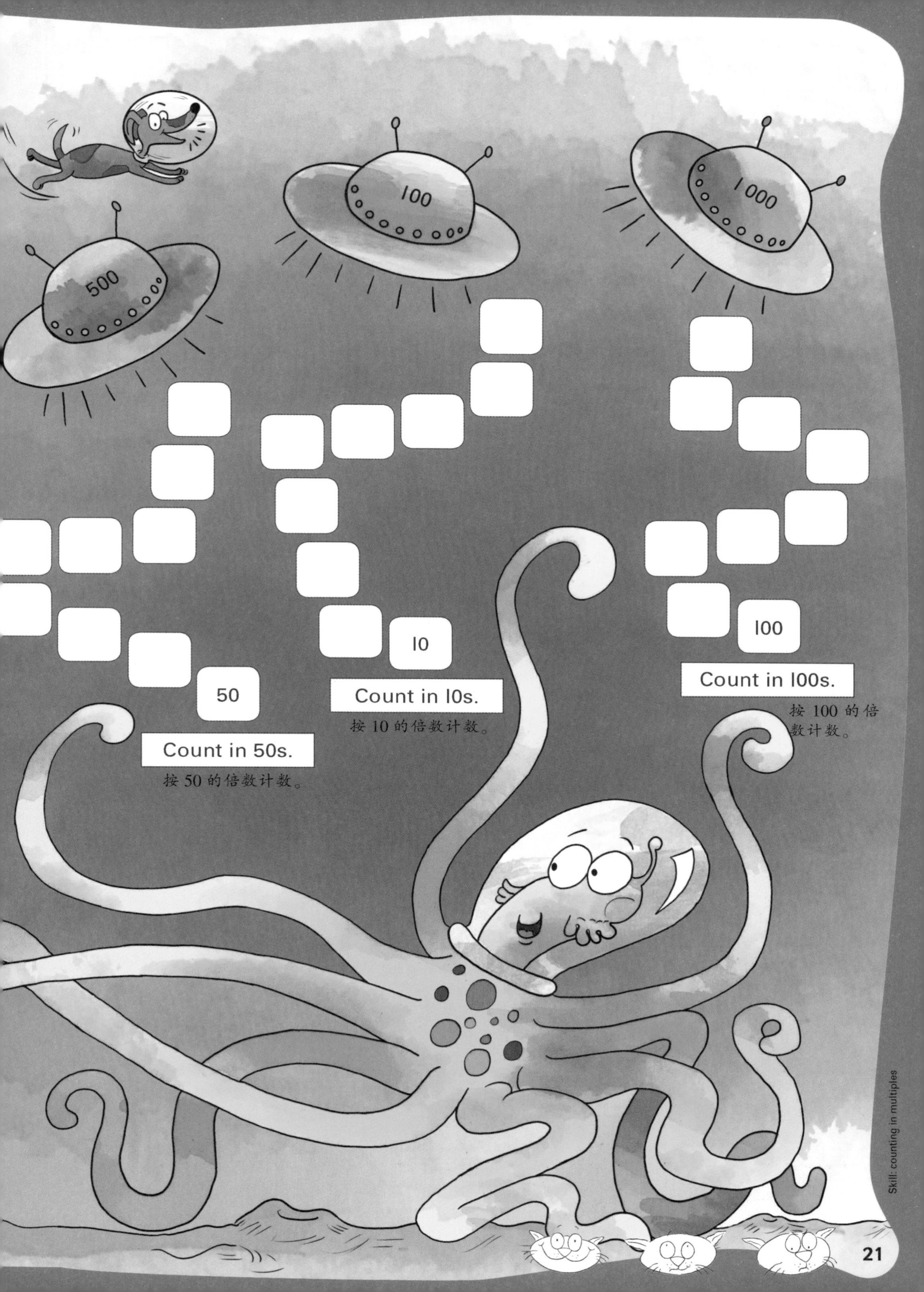
500
100
1000
50
Count in 50s.
按 50 的倍数计数。
10
Count in 10s.
按 10 的倍数计数。
100
Count in 100s.
按 100 的倍数计数。

Space repairs

太空维修

The spaceships need to be repaired but the naughty aliens have muddled up the spare parts. Can you find which parts belong to each spaceship?

宇宙飞船需要修理,但调皮的外星人将备件弄乱了。你能找出哪些备件属于哪只宇宙飞船的吗?

1. All the parts with odd numbers belong to the Goops. All the parts with even numbers belong to the Zoops. Draw red lines to the Goops and blue lines to the Zoops to show this.
   所有的奇数备件属于 Goops。所有的偶数备件属于 Zoops。像下面这样用红色线连 Goops,蓝色线连接 Zoops。
2. Break up each of the numbers into tens and units.
   将每个数字折分成十位和个位。

Skills: decomposing numbers, identifying odd and even numbers

Bee prepared!

让蜜蜂做好准备！

Bees work very hard in spring! They have been so busy that they have forgotten the order in which to leave the hive. The bee with the smallest number should leave the hive first.

在春天蜜蜂工作非常辛苦！它们很忙，以至于忘了离开蜂巢的顺序。数字最小的蜜蜂应该最先离开蜂巢。

1. Look at the numbers on the honeycomb. 看蜂巢上的数字。
2. Stick a number on each bee so that the honeycomb numbers are in order from smallest to biggest. Start with the bee with the smallest number.
   按照蜂巢上的数字从小到大的顺序给每只蜜蜂贴上编号。从编号最小的蜜蜂开始。

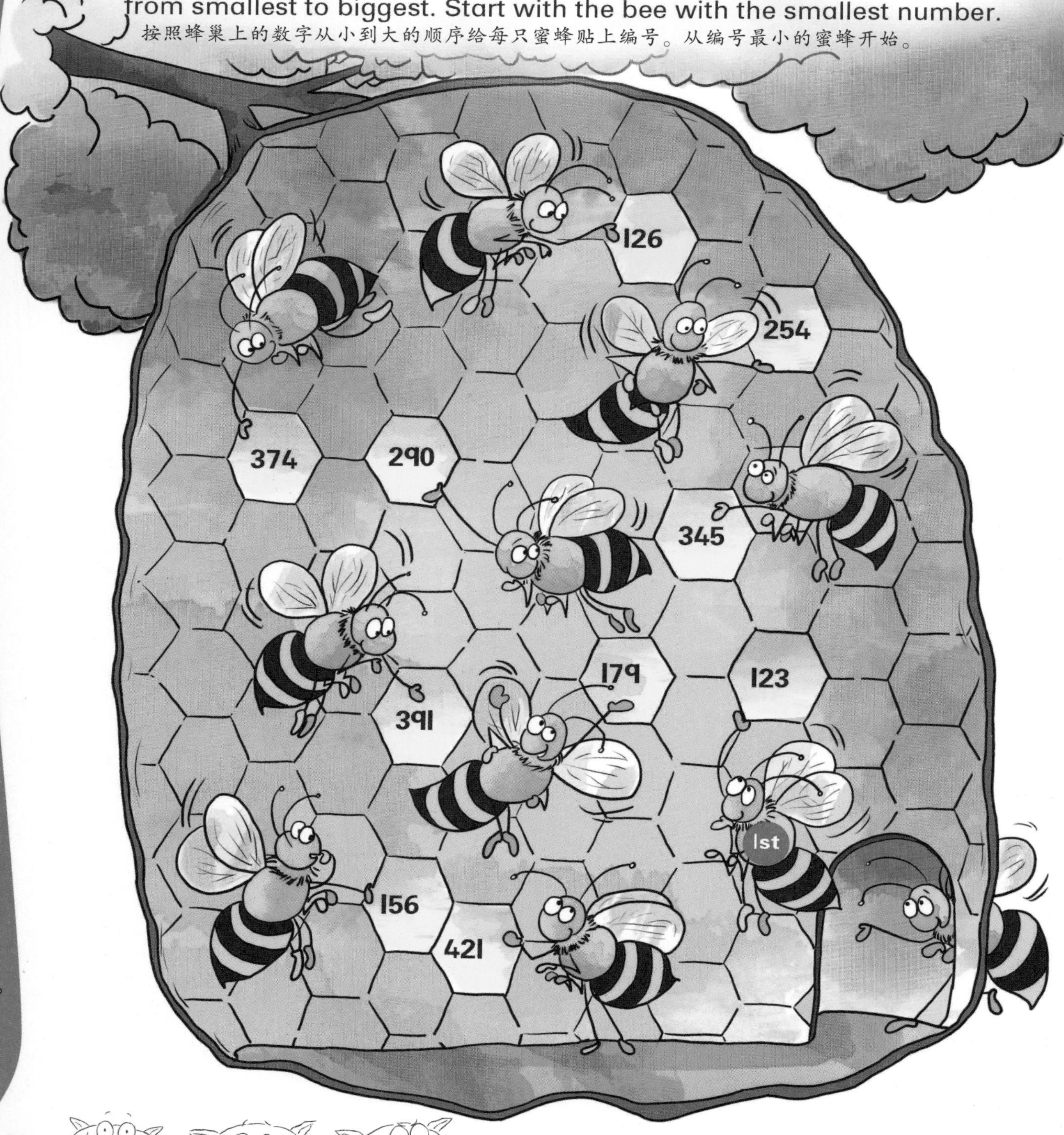

What do you like? 你喜欢什么？

Some people like books or DVDs that give information.
Other people like books or DVDs that tell stories.
What do you and your friends or family like best?
有些人喜欢通过看书或者光盘来了解信息。有些人喜欢通过书或者光盘来讲故事。你和你的朋友或家人最喜欢什么？

1. Write your name in the first block. Colour in the blocks in your row to show what you like best.
   在第一个方框中写上你的名字。在你这一行里给你最喜欢的那一类所对应的方框涂上颜色。
2. Now do this for five other people. 现在需要了解另外5个人。

	Name 名称	Information 信息	Story 故事	Book 书	DVD 光盘
1					
2					
3					
4					
5					
6					

3. Answer the questions.
   回答这些问题。
   a. Which is the most popular kind of book or DVD? ____________________
      哪种书或光盘最受欢迎？
   b. Which is the least popular kind of book or DVD? ____________________
      哪种书或光盘最不受欢迎？

Skill: working with graphs

Heavy or light? 重或轻？

One way to decide whether objects are heavy or light is by holding them in your hands. 确定物品轻重的一种方法是把它们放在你的手上。

1. Collect the pairs of objects below.
   收集下面物品。
2. Hold one in each hand.
   每只手拿起一样。
3. Decide which one feels heavy and which one feels light.
   用感觉来确定哪个重，哪个轻。
4. Write heavy or light under each picture.
   在每个图下面写出重或轻。

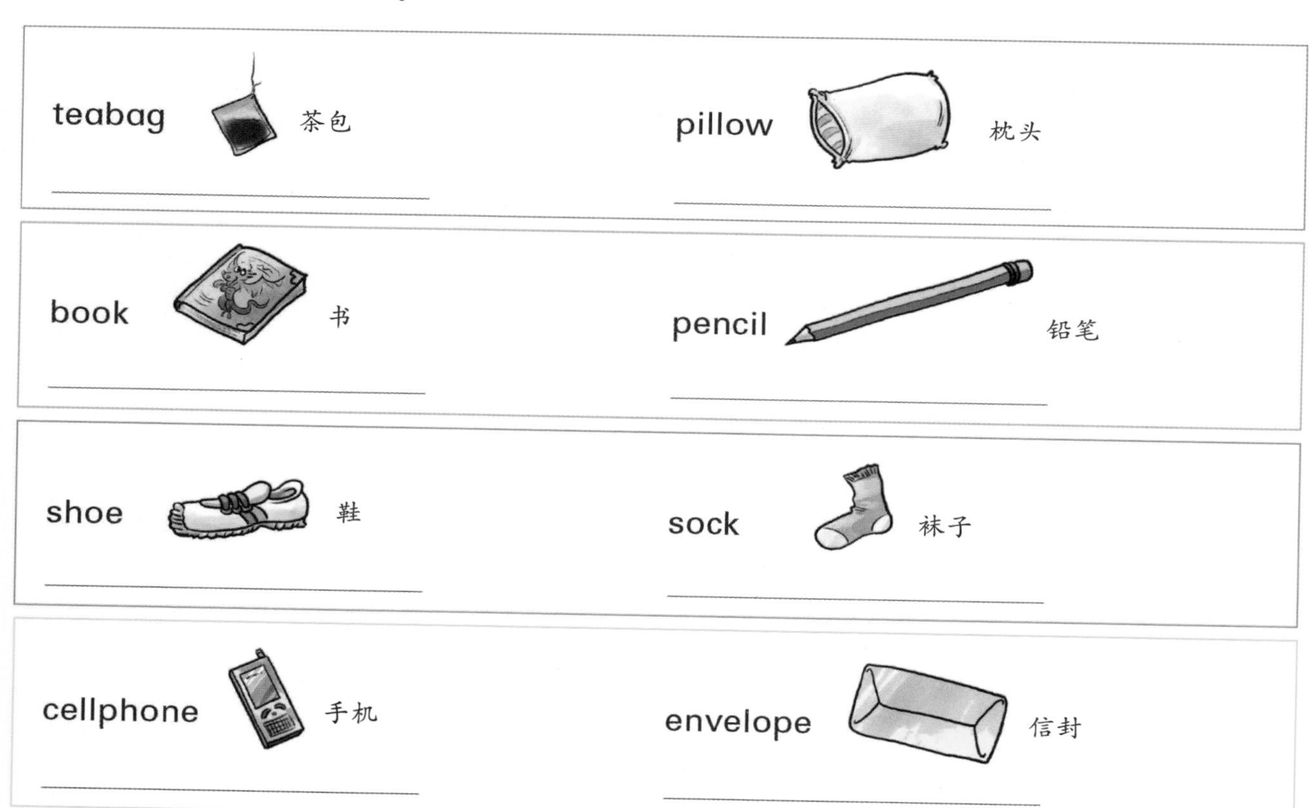

Term 2

Skill: comparing mass

More busy bees

更多忙碌的蜜蜂

Help the busy bees number their honeycombs.

帮助忙碌的蜜蜂填写蜂巢中的数字。

Break up each number into three parts: hundreds, tens and units.

将每个数分成三部分：百位、十位和个位。

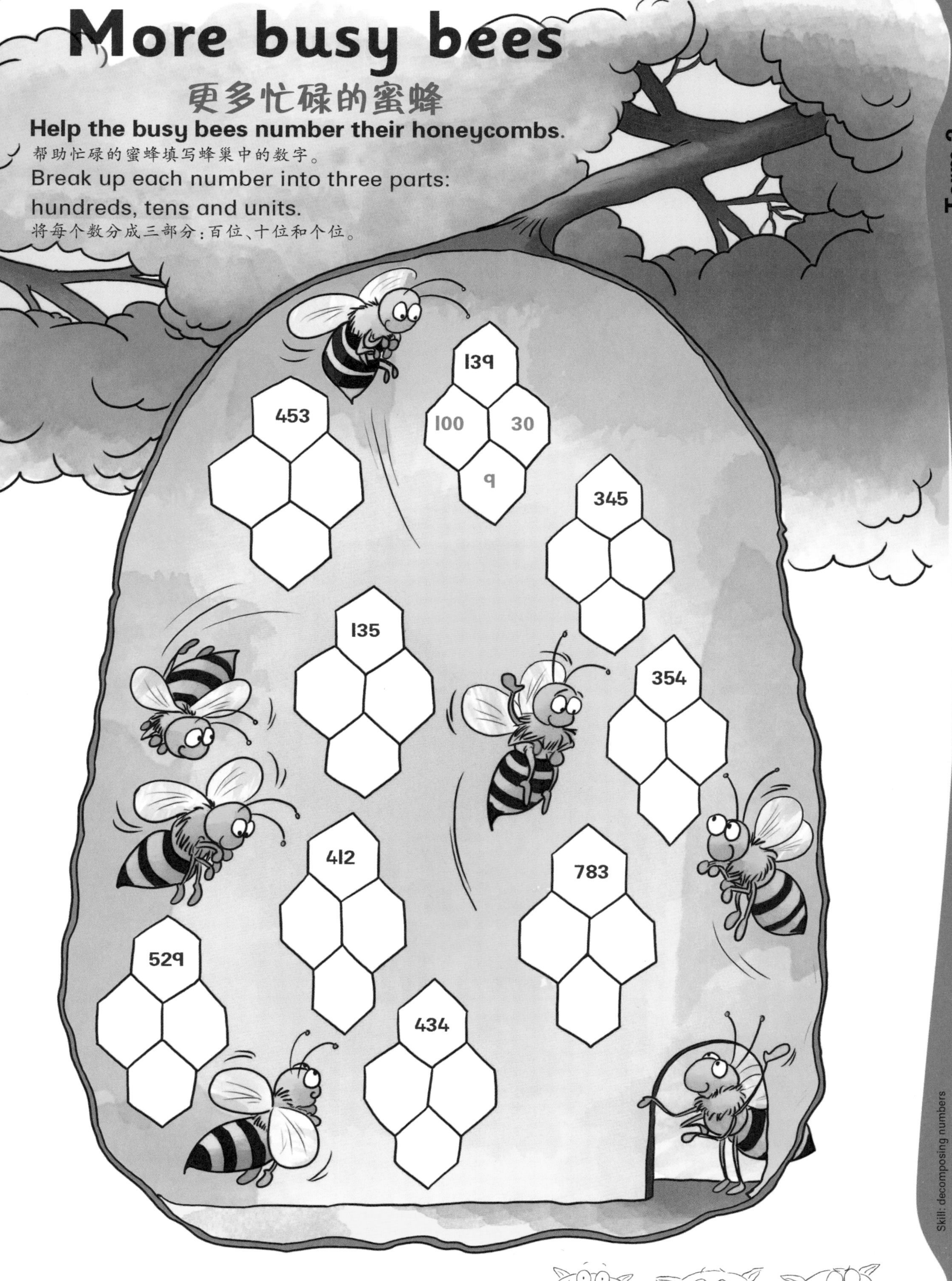

Building houses 盖房子

The three little pigs are building their houses. Can you help them work out what they need?

三只小猪正在建造它们的房子。你能帮助它们算出需要什么吗？

1. This little pig is building a house of straw. Draw a line of straw to join the two number sentences that have the same answer.
   这只小猪正在建一座稻草房，将两组答案相同的秸秆连线。

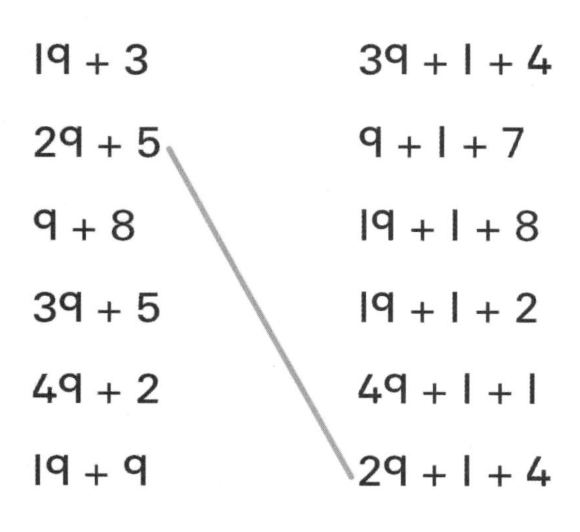

19 + 3	39 + 1 + 4
29 + 5	9 + 1 + 7
9 + 8	19 + 1 + 8
39 + 5	19 + 1 + 2
49 + 2	49 + 1 + 1
19 + 9	29 + 1 + 4

2. This little pig is building a house of sticks. Complete the number sentences.
   这只小猪正在建一座板条房。完成这些算术题。

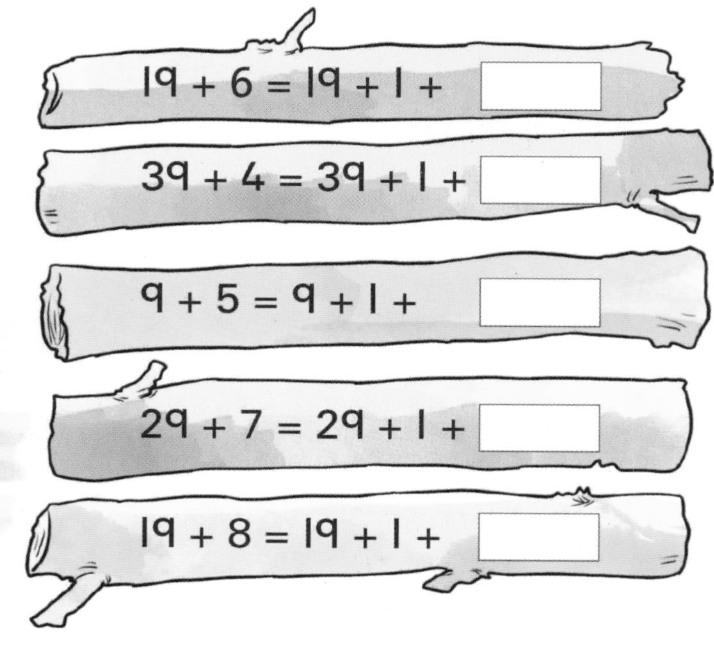

19 + 6 = 19 + 1 + ☐

39 + 4 = 39 + 1 + ☐

9 + 5 = 9 + 1 + ☐

29 + 7 = 29 + 1 + ☐

19 + 8 = 19 + 1 + ☐

3. Can you help this little pig build a house of bricks? Complete the sums in the bricks.
你可以帮助这只小猪建造一座砖房子吗？请算出砖块的总数。

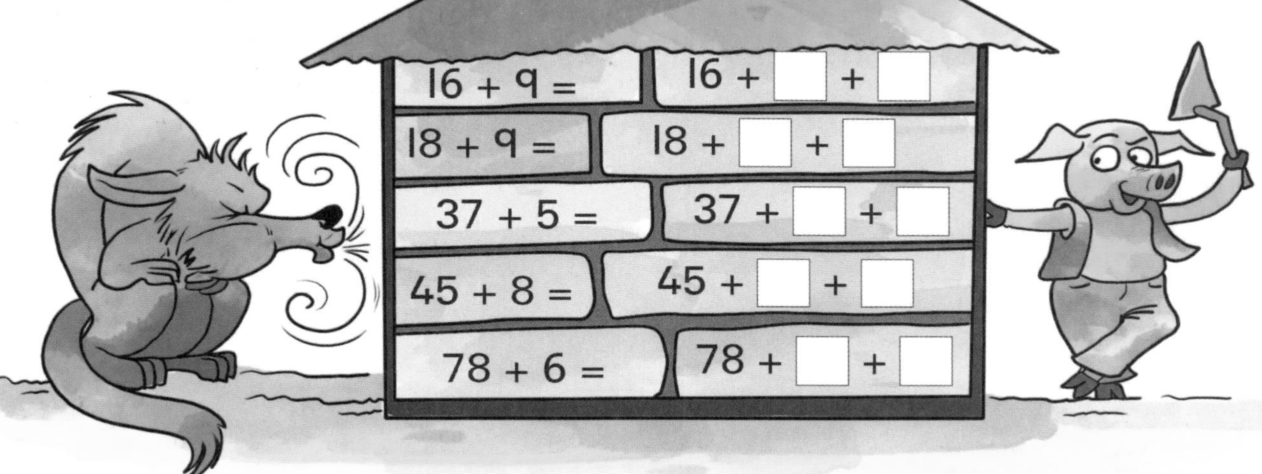

4. Choose one brick from each pile and add the numbers together to make the numbers on the cement bags. Use each number once. Look at the example. 分别从每堆选出 1 块砖，将它们上面的数字加在一起使其等于水泥袋上的数字。每堆数字只能用 1 次，看一看例子。

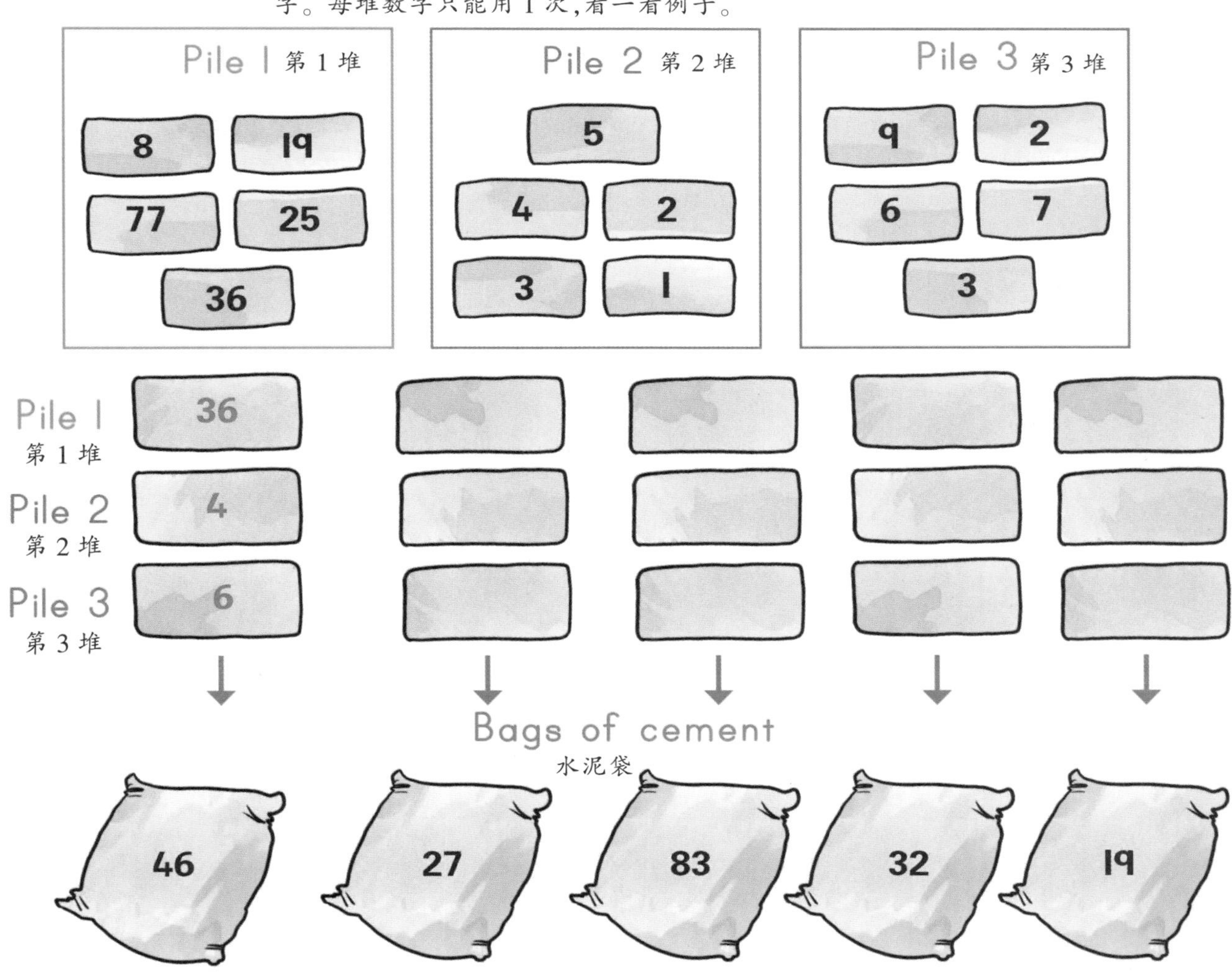

Skill: addition

Magic shapes 神奇的图形

Parts of these shapes and patterns have disappeared.
这些图形和图案的一部分消失了。

1. Complete the symmetrical shapes.
   完成对称图形。
2. Which of the shapes have straight sides?
   Colour the shapes green.
   哪些图形有直线边？将图形涂成绿色。
3. Which of the shapes have round sides? 哪些图形有圆边？
   Colour the shapes yellow. 将图形涂成黄色。

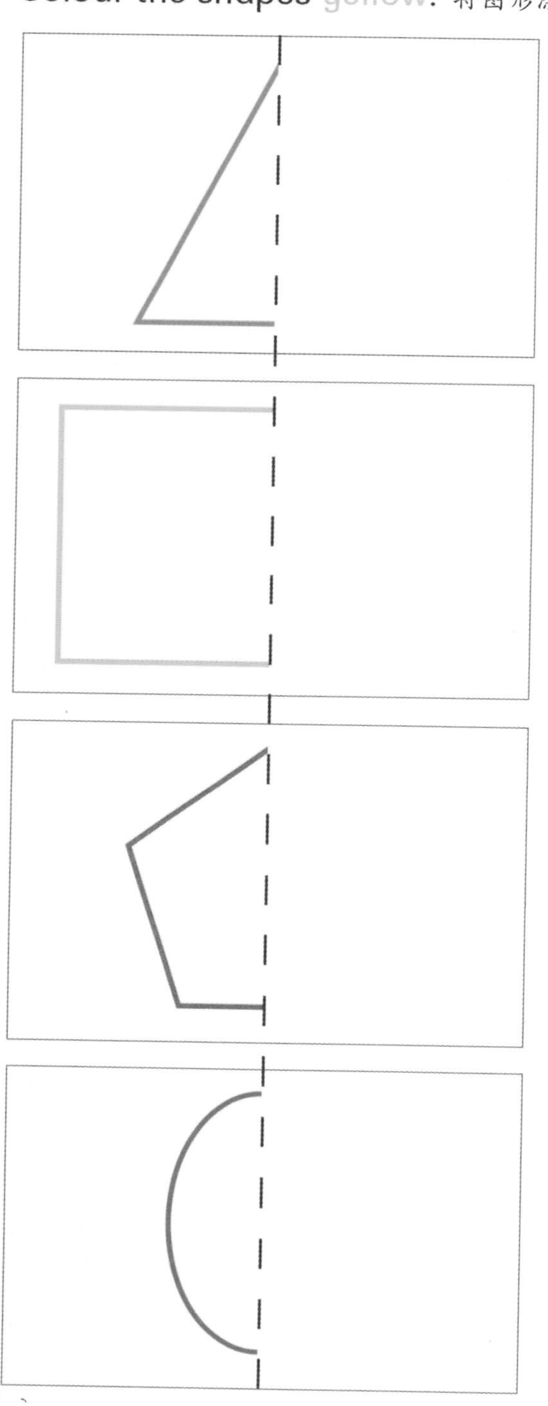

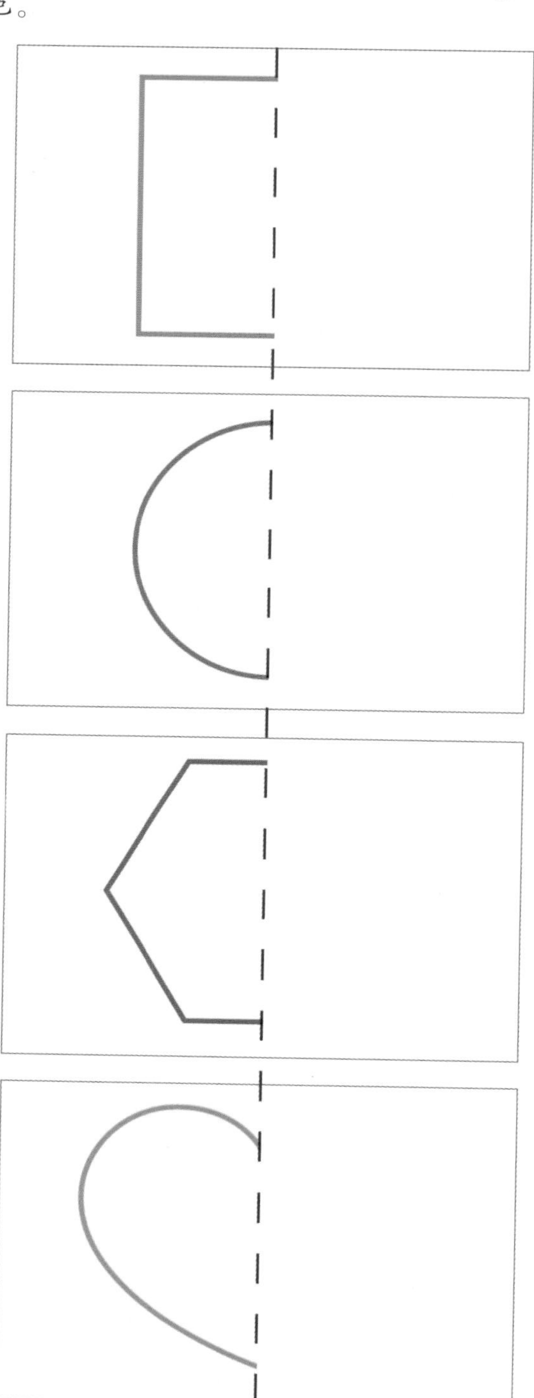

How short is short? 量量有多短？

Have you ever used a ruler to measure? 你用尺测量过吗？

Find a pencil, peg, soap, toothpick, eraser and key to measure.

找出铅笔、衣夹、肥皂、牙签、橡皮和钥匙用来测量。

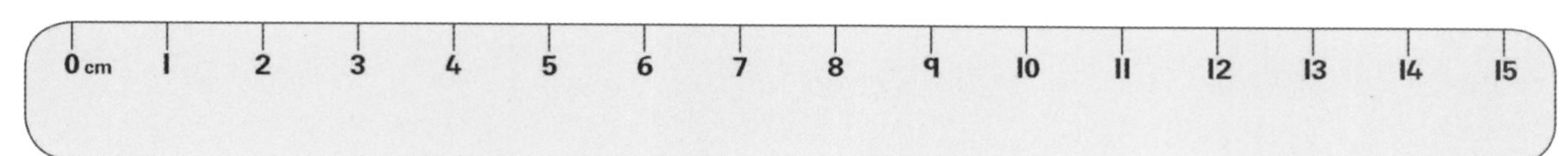

1. Estimate how many centimetres long each object is. Record this number in the table.
估计每个物体的长度是多少厘米。将数字记录在下面表格中。
2. Measure each object using a ruler. Record this number in the table.
用尺来测量每个物体。将数字记录在下面表格中。

	Estimate 估计值	Measure 测量值

	Estimate 估计值	Measure 测量值

3. Which of your estimates are the same as your measurements?
你的哪个估计值与测量值相同？
____
4. Which was your closest estimate? _______________________
哪个估计值最接近？
5. Are any objects the same length? _______________________
长度相同的物体是哪些？
6. Write the names of the objects in order from **shortest** to **longest**.
按照从短到长的顺序，写下这些物体的名称。
____
____
____
____
____
____

Term 2

Skill: measuring length

Monkey business

恶作剧

Monkeys love getting up to mischief. This monkey is trying to steal bunches of bananas!

猴子们喜欢搞一些恶作剧。
这只猴子试图偷一把香蕉！

1. Colour the bunch that you think (estimate) will add up to the highest number.
   将你认为(估计)加起来数字最大的那把香蕉涂色。

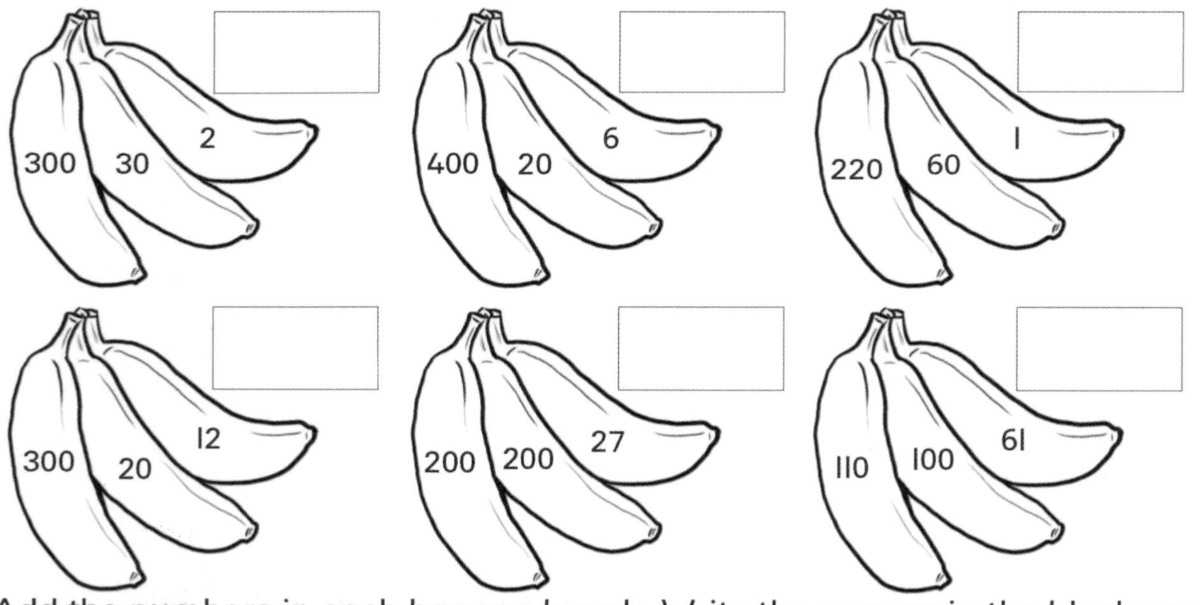

2. Add the numbers in each banana bunch. Write the answer in the block next to each bunch. 将每把香蕉上的数字相加。将答案写在每把香蕉邻近的方框中。
3. Which bunches add up to the same number? Colour the answer blocks the same colour. 哪几把香蕉上相加的数字相同？将答案相同的方框涂上同一种颜色。
4. Did you estimate and colour the correct bunch? If not, now colour in the bunch which adds up to the highest number.
   你估计并涂色的那把香蕉对了吗？如果不对，现在将那把数字相加最高的香蕉涂色。

Skill: addition

How long is long?

量量有多长？

Can you tell how long something is by looking at it?
你能一看就知道物品有多长吗？

1. Estimate. 估计值。
   - Which of these things do you think is the **longest**?
     你认为这些物品中哪个最长？
   - Which of these things do you think is the **shortest**?
     你认为这些物品中哪个最短？

bedroom
卧室

bath
浴缸

car
汽车

bed
床

window
窗户

chair
椅子

2. Write them in order from **longest** to **shortest**. 按照从长到短的顺序把它们写下来。

   ____________, ____________, ____________,

   ____________, ____________, ____________

3. Using your footsteps, measure how long each thing is in real life.
   Record your answers. 用你的脚步测量一下生活中的每件物品有多长。记录你的答案。

4. Using a 1 m piece of string, measure how long each thing is.
   Record your answers. 用一根1米长的绳子测量一下每件物品有多长。记录你的答案。

	Footsteps 脚步	Metres 米
The bedroom is 卧室是		
The window is 窗户是		
The bed is 床是		
The car is 汽车是		
The bath is 浴缸是		
The chair is 椅子是		

5. Which is the **longest**? ____________
   哪个最长？

6. Which is the **shortest**? ____________
   哪个最短？

Skill: measuring length

Classroom design 教室布局

Here is a map of Ravi's classroom. Look carefully at it and answer the questions on page 35.

下图是 Ravi 的教室平面图。仔细观察并回答第 35 页的问题。

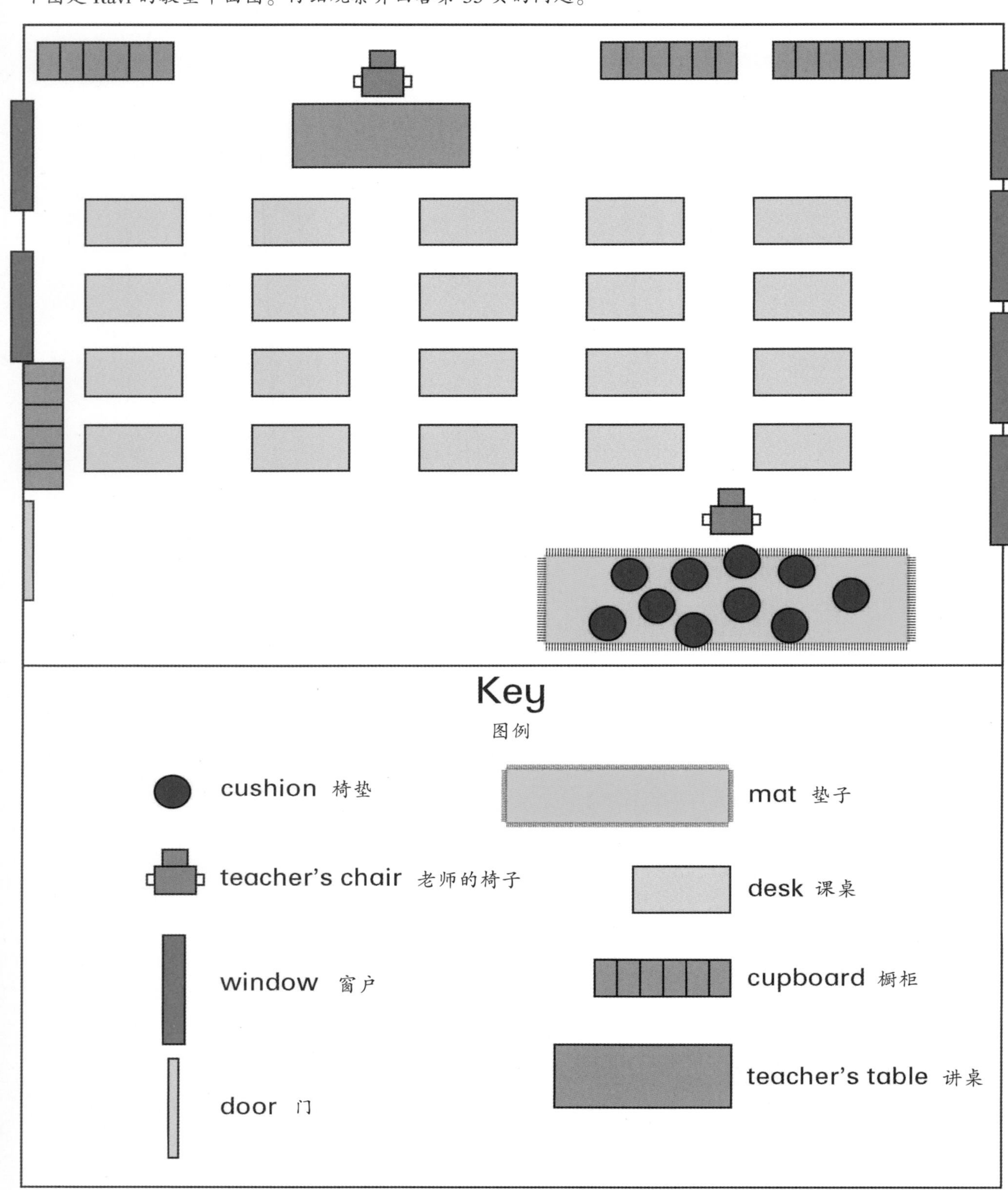

1. How many desks are in the classroom?
教室里有多少张课桌？

2. How many windows can you see?
你能看见多少扇窗户？

3. How many cushions are there?
有多少个椅垫？

4. There are 20 children in Ravi's class. Is there a cushion for each child? ______
在 Ravi 的班级里有 20 个孩子。每个孩子都有一个椅垫吗？

5. How many more does the class need?
这个班级还需要多少个椅垫？

6. How many cupboards are in this classroom?
这间教室里有多少个橱柜？

7. What is nearest to the door? ______
最靠近门的是什么？

8. Where is the teacher's table? ______
老师的讲桌在哪儿？

9. Where are the teacher's chairs? ______
老师的椅子在哪儿？

10. How many desks are in each row?
每排有多少张课桌？

11. Draw a symbol for a bin below. Add a bin to the map and to the key.
在下面画一个箱子的符号。在平面图和图例上增加一个箱子。

12. Add two cushions to the mat. How many cushions are there altogether?
在垫子上加 2 个椅垫。现在总共有多少个椅垫？

13. Draw a symbol for a blackboard below. Add a blackboard to the map and to the key. 在下面画一个黑板的符号。在平面图和图例上增加一个黑板。

Catch a fraction

捕捉分数

Pirate Jody has thrown her bottles overboard.
海盗 Jody 把她的瓶子扔到船外。

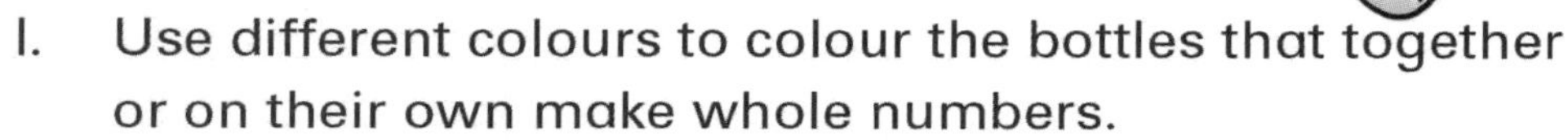

1. Use different colours to colour the bottles that together or on their own make whole numbers.
将那些分数加在一起变成整数或者本身就是整数的瓶子涂上不同的颜色。

2. How many whole numbers did you make? ☐
你能组合出多少整数？

3. What was left over? ☐
剩下的是什么？

Weigh to go! 称重！

How much do these things weigh?

这些物品的重量是多少？

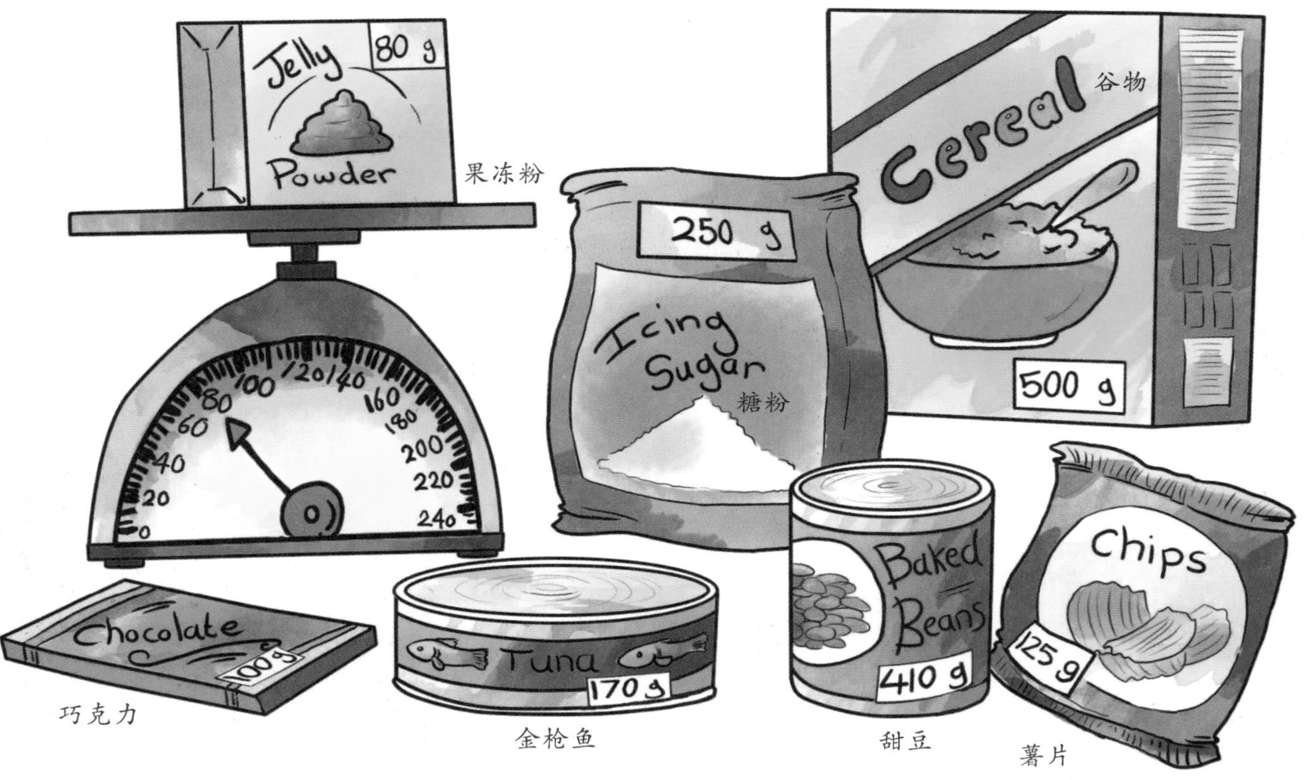

1. Work out how much these things weigh.
   计算出这些物品的重量。

   a. 100 g chocolate + 170 g tuna = ☐ g
   100 克巧克力+170 克金枪鱼。

   b. 80 g jelly powder + 410 g baked beans = ☐ g
   80 克果冻粉+410 克甜豆。

   c. 125 g chips + 170 g tuna = ☐ g
   125 克薯片+170 克金枪鱼。

   d. Which two things weigh the most? ________________ ________________
   哪两件物品最重？

2. What weighs the same as two packets of icing sugar? ________________
   与两袋糖粉重量相同的是什么？

3. How many slabs of chocolate would weigh the same as the box of cereal? ☐
   几板巧克力和这盒谷物的重量相同？

4. How many packets of chips would weigh the same as a packet of icing sugar? ☐
   几袋薯片和 1 袋糖粉的重量相同？

Jungle soccer fever
丛林足球热

Which team is the best jungle soccer team of the season?
哪支球队是本赛季最好的丛林足球队？

1. Count up each team's points.
   计算出每队的得分。

2. Answer the questions. 回答问题。

a. Which team scored the most points this season? 本赛季哪支球队得分最多？

____

b. Which team scored the fewest points? 哪支球队得分最少？

____

c. How many teams scored more than 35 points? ☐ 有多少支球队得分超过 35 分？

d. Which teams scored the same number of points?
哪些球队的得分相同？

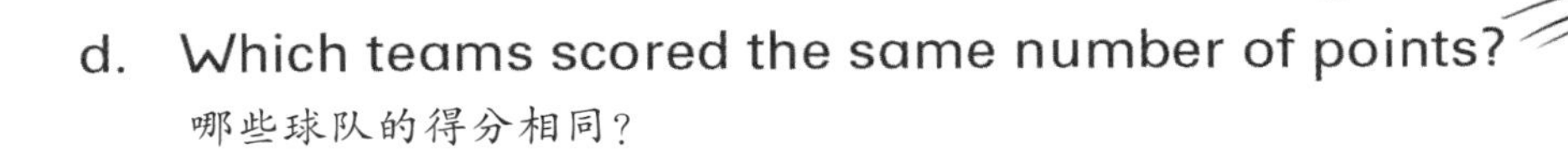

____

e. How many teams have even numbers as their total scores? ☐ 有多少支球队得分是偶数？

f. Which team scored 30 points more than the Singing Sloths? ______________________________
哪支球队得分比唱歌的树獭多 30 分？

____

Skill: addition

Leapfrog 蛙跳

The river frogs are having a jumping competition. The boys are playing high jump.
河里的青蛙在举行一场跳跃比赛。公青蛙们正在跳高。

1. Double the numbers on the side of the high jump poles to find out how high the boy frogs jumped. Look at the example.
将跳高杆右边的数字加倍,看这些公青蛙跳了多高。看下面例子。

Which boy frog jumped the highest? ____________________
哪一只公青蛙跳得最高?

The girl frogs are also playing high jump.

母青蛙们也玩起了跳高比赛。

2. Halve the numbers on the side of the high jump poles to find out what numbers the girl frogs landed on. Look at the example.

将跳高杆右边的数字减半,以便找出母青蛙落到了哪个数字。看下面例子。

Which girl frog jumped the lowest? ______________________

哪一只母青蛙跳得最低?

Sunday sundae

圣代冰淇淋

RMB 5yuan =500 c

You have 5yuan to spend on an ice-cream sundae.
Think carefully about the toppings you would like to buy.
你有 5 元可以用来买圣代冰淇淋。认真考虑好，你最想买什么配料的。

1. Rewrite the prices in yuan.
按人民币改写价格。

caramel chips 焦糖	yuan	ice-cream scoop 冰淇淋勺	yuan
fudge 软糖	yuan	chocolate sauce 巧克力酱	yuan
chocolate sprinkles 巧克力碎粒	yuan	strawberries 草莓	yuan

2. Write down your order. Remember that you only have 5yuan!
写下你的订单。记住你只有 5 元。

Order 订单

____

____

____

Total: RMB ____________
总计：

3. Draw your ice-cream sundae in the glass below.
在下面杯子画出你的圣代冰淇淋。

4. Here are some Special Sundaes on the menu. Write the prices in yuan.
菜单上有一些特色圣代,写出以元为单位的价格。

Special Sundaes 特色圣代		
	cents 分	yuan 元
Chocolate Madness 疯狂巧克力	340c	
Banana Jam 香蕉果酱	270c	
Strawberry Supreme 至尊草莓	490c	
Caramel Craze 狂热焦糖	610c	
Vanilla Dream 香草之梦	345c	
Berry Delicious 美味果仁	295c	

5. Which Special Sundae is the most expensive? ______________________
哪种特色圣代最贵?

6. Which Special Sundae is the cheapest? ______________________
哪种特色圣代最便宜?

Sweet treats 甜食

1. Can you find four different types of sweets on the tray?
你能在托盘上找出 4 种不同类型的糖果吗？

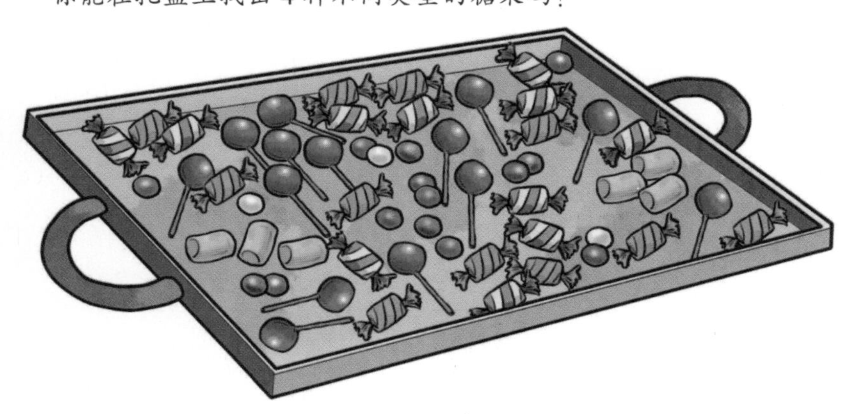

I can see: 我能看到：

lollipops
棒棒糖

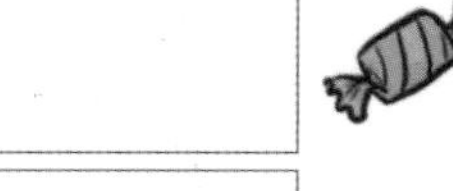

sweets
糖果

jelly beans
软糖豆

marshmallows
果汁软糖

2. Help Emma to share these sweets between different groups of children. How many does each child get? Write it in the first block. How many are left? Write it in the second block.
帮助 Emma 给不同组的孩子分糖果。每个孩子分到多少？写在第 1 个方框里。剩下的是多少？写在第 2 个方框中。

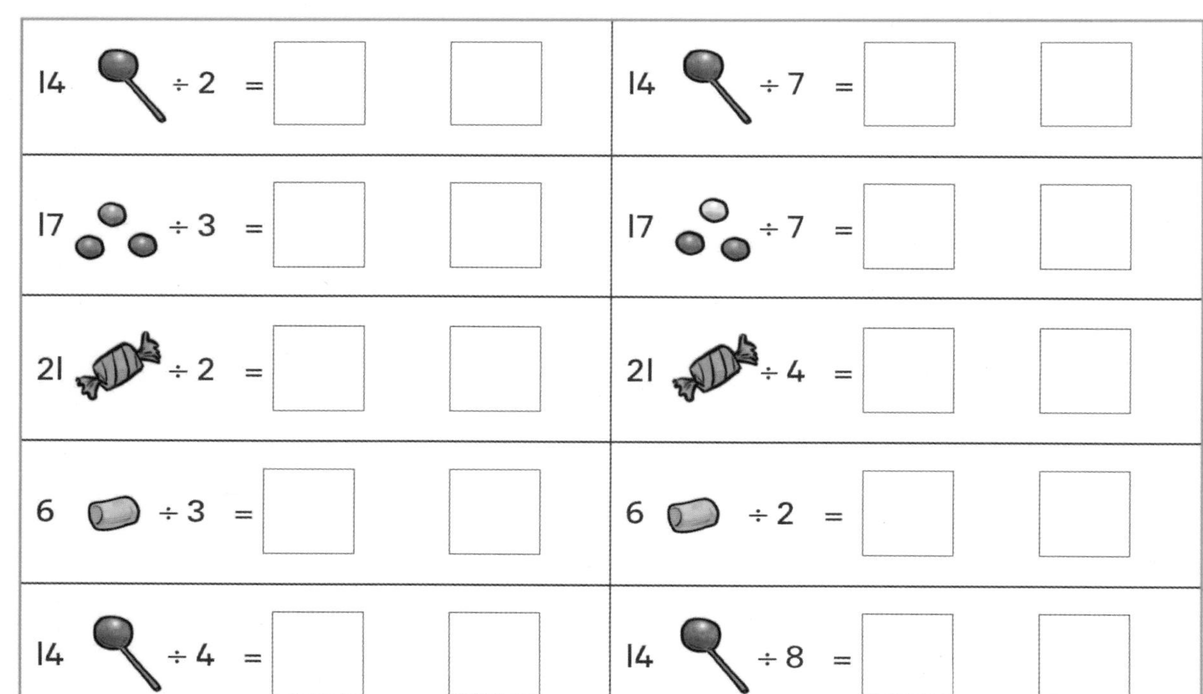

14 lollipops ÷ 2 =	☐	☐	14 lollipops ÷ 7 =	☐	☐
17 jelly beans ÷ 3 =	☐	☐	17 jelly beans ÷ 7 =	☐	☐
21 sweets ÷ 2 =	☐	☐	21 sweets ÷ 4 =	☐	☐
6 marshmallows ÷ 3 =	☐	☐	6 marshmallows ÷ 2 =	☐	☐
14 lollipops ÷ 4 =	☐	☐	14 lollipops ÷ 8 =	☐	☐

Skill: division

Good timing 分秒不差

Different types of clocks display the time in different ways. The clocks on this page are called digital clocks.

不同类型的时钟时间显示的方式不同。
本页的时钟叫数字时钟。

1. Say the time on each clock. 说出每个时钟显示的时间。
2. Write it down. 写下来。

Twenty minutes past seven in the morning

21:08

11:45

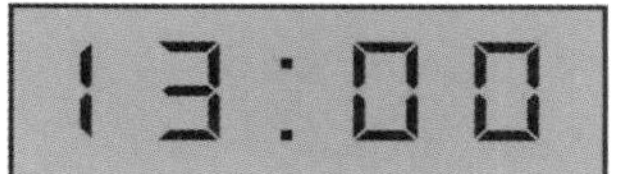

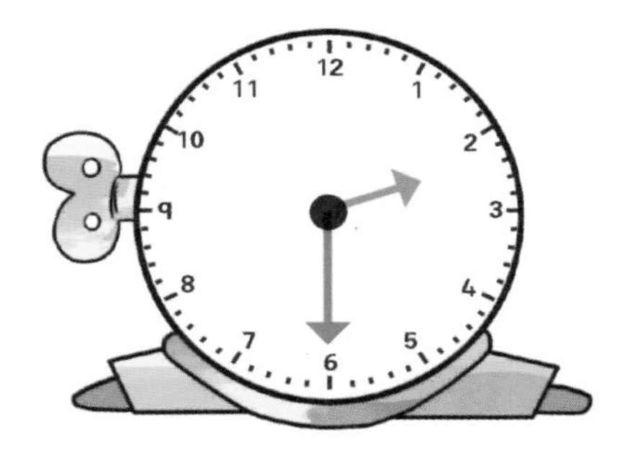

3. Draw the hands on the clock to show the time.
在时钟上画出指针。

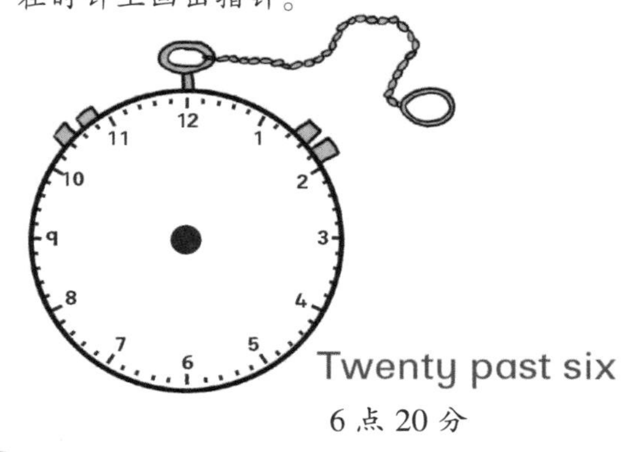

Twenty past six
6 点 20 分

4. Write the time on this clock:
写出这个时钟所示时间：

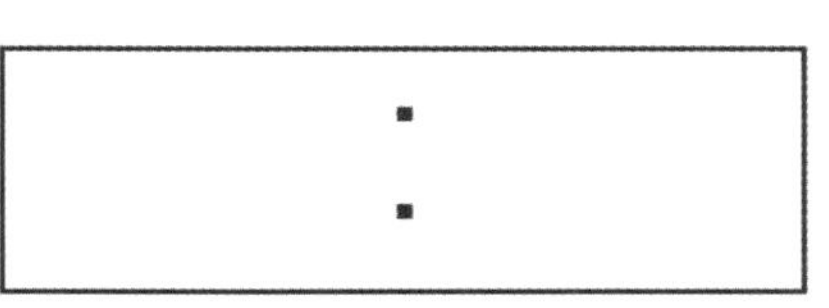

Twenty-five minutes past seven in the morning
上午 7 点 25 分

Term 3

Pizza pizzazz

美味披萨

Lebo is having a pizza party. Lebo asked her friends which two toppings they each like best. Lebo added up the numbers for each topping and wrote this down.

Lebo 正举办一个披萨晚会。她让她的朋友每人都选出两种最喜欢的配料。Lebo 把每种配料喜欢的人数加起来，并记录下来。

Topping 配料	
Salami 意大利香肠	I I I I I I I I = 8
Olives 橄榄	I = I
Pineapple 菠萝	I I I I I I = 6
Mushrooms 蘑菇	I I I = 3
Chicken 鸡肉	I I I I = 4
Extra cheese 特色奶酪	I I = 2

I. Complete the graph for Lebo. Colour each row a different colour.
帮 Lebo 完成这个图表。将每一行涂上不同的颜色。

Favourite pizza toppings 最喜欢的披萨配料						
8						
7						
6						
5						
4						
3						
2						
I						
	Salami 意大利香肠	Olives 橄榄	Pineapple 菠萝	Mushrooms 蘑菇	Chicken 鸡肉	Extra cheese 特色奶酪

2. Answer the following questions: 回答下列问题：
   a. How many children took part in Lebo's survey? ☐
      有多少孩子参加了 Lebo 的调查？
   b. What was the favourite topping? ____________________
      最喜欢的配料是什么？
   c. What was the least favourite topping? ____________________
      最不喜欢的配料是什么？
   d. How many more children preferred salami to pineapple?
      Complete the sentence: 喜欢意大利香肠的孩子比喜欢菠萝的孩子多几个？
      完成这个句子：
      ☐ – ☐ = ☐ more children preferred salami.
      多___个孩子更喜欢意大利香肠。
   e. How many more children preferred chicken to extra cheese?
      Complete the sentence: 喜欢鸡肉的孩子比喜欢特色奶酪的孩子多几个？
      完成这个句子：
      ☐ – ☐ = ☐ more children preferred chicken.
      多___个孩子更喜欢鸡肉。
   f. Lebo's mother said that she could only afford to buy four types of toppings. Which should she choose? Why?
      Lebo 妈妈说，她只能购买 4 种配料。她应该选择哪几种？为什么？

      ____

      ____

      ____

      ____

3. Choose toppings from the sticker page and make your pizza here.
   从贴纸页面选出这几种配料并在下面做出你的披萨。

Skill: working with graphs

Puzzling perimeter

周长之谜

Do you know what perimeter is? Perimeter is a measurement. When you measure around the outside of an object you are measuring its perimeter.

你知道什么是周长吗？周长是一种测量值。
当你环绕物体周边测量时，你测量的就是它的周长。

Here is a drawing of a soccer field. The red line around the outside shows you the perimeter of the soccer field.

这是一个足球场的平面图。外周的红线标出了足球场的周长。

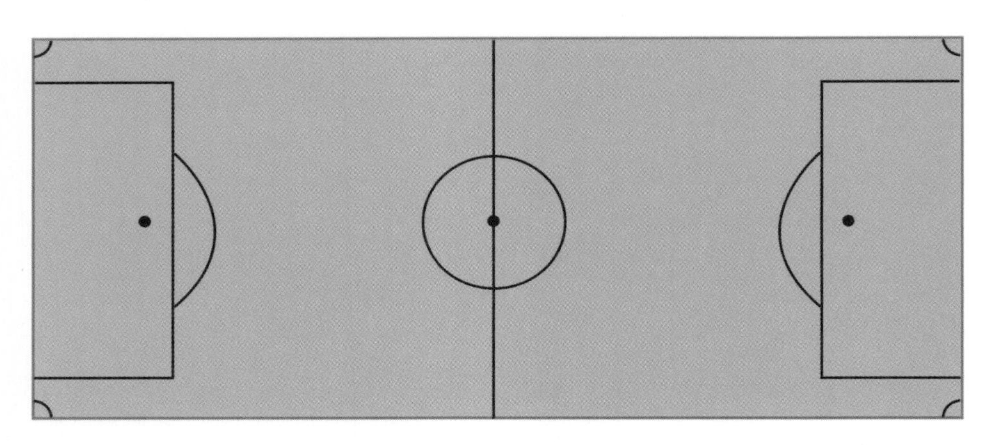

Now try measuring the perimeter of a box. 现在试着测量一个盒子的周长。

Find: string, sticky tape, scissors, a small box. 找出：绳子，胶带，剪刀，小盒子。

1. Stick the end of a piece of string to one corner of your box.
   用胶带将绳子的一端贴在盒子的一个角。
2. Run the string along the sides of your box until you get back to where you started.
   沿着盒子边缘伸展绳子直至返回到开始的部位。
3. Cut the end of string exactly where it meets the beginning. 将绳子正好在与其开始部位会合处剪断。

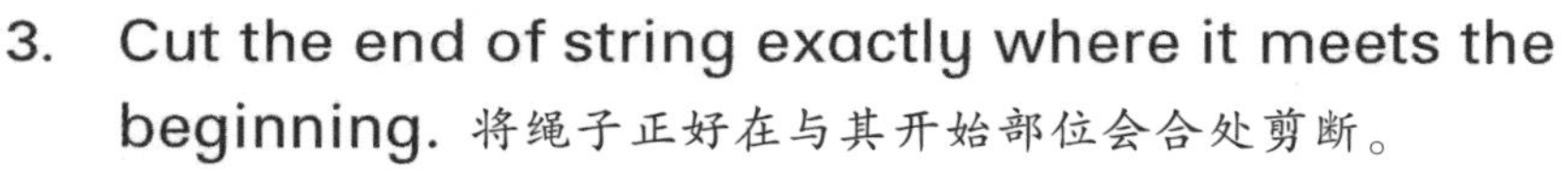

You have measured the perimeter of your box!
你就测量出盒子的周长了！

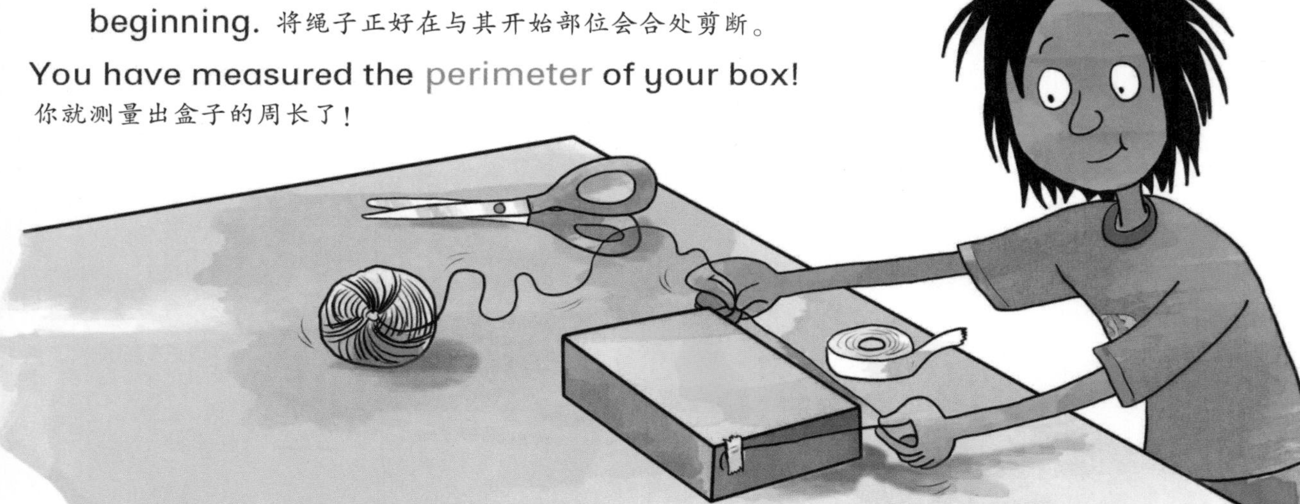

Matching cards 匹配卡片

These cards belong to the same pack.
这些卡片属于同一副。

1. Complete the sums on each card. Write the answer at the bottom of the card.
   完成每张卡片的算术题。将答案写在卡片底部。
2. Colour the cards with the same answers in the same colour.
   将答案相同的卡片涂上相同颜色。

3. The smallest number is ______.
   最小的数字是______。

Fraction action

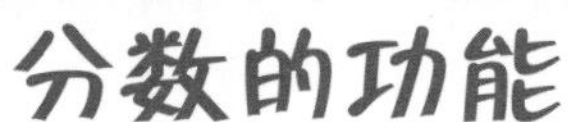

Mandla, Ravi, Jody, Lebo, Jaco and Emma have enjoyed their morning at the school fair. They each bought a pizza and a bag of nuts to enjoy in the car on their way home. Mandla、Ravi、Jody、Lebo、Jaco 和 Emma 都喜欢在早晨去学校的市场。他们每人买了一个披萨和一袋坚果准备在回家路上的车里享用。

1. Use your stickers to show the amount of pizza each child ate:
   用贴纸表示出每个孩子吃披萨的数量：
   a. Jody ate half of her pizza. Jody 吃了 $\frac{1}{2}$ 的披萨。
   b. Mandla ate three quarters of his pizza. Mandla 吃了 $\frac{3}{4}$ 的披萨。
   c. Jaco ate a whole pizza. Jaco 吃了整个披萨。
   d. Ravi ate three-eighths of his pizza. Ravi 吃了 $\frac{3}{8}$ 的披萨。
   e. Lebo ate six-eighths of her pizza. Lebo 吃了 $\frac{6}{8}$ 的披萨。
   f. Emma only ate a quarter of her pizza. Emma 只吃了 $\frac{1}{4}$ 的披萨。
2. Then write the fraction under each pizza.
   然后在每个披萨上写出分数。

a. Jody

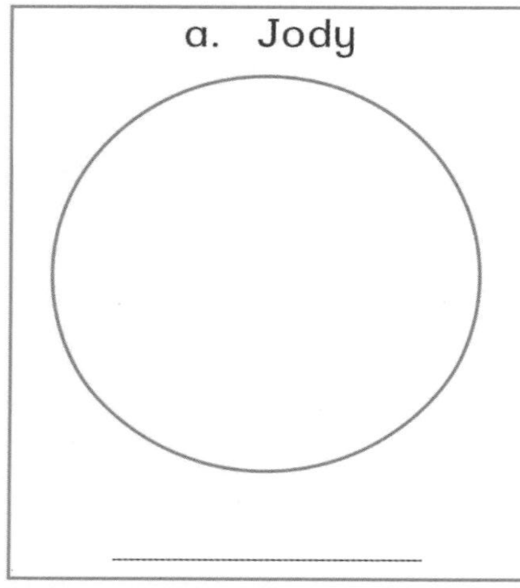

b. Mandla

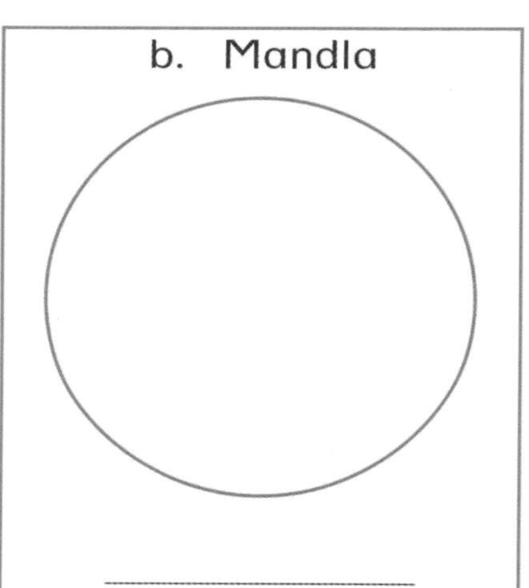

c. Jaco

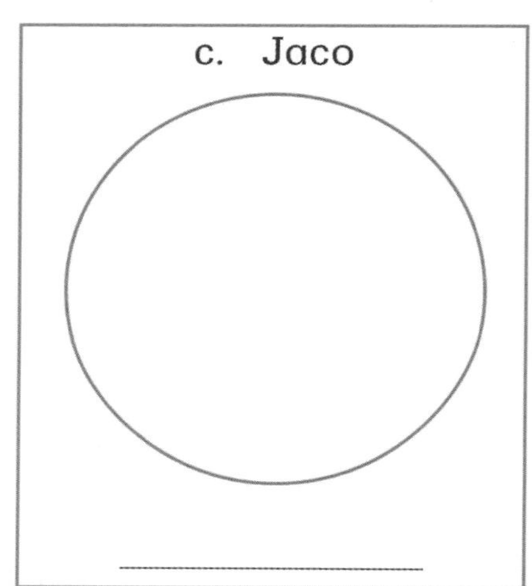

d. Ravi

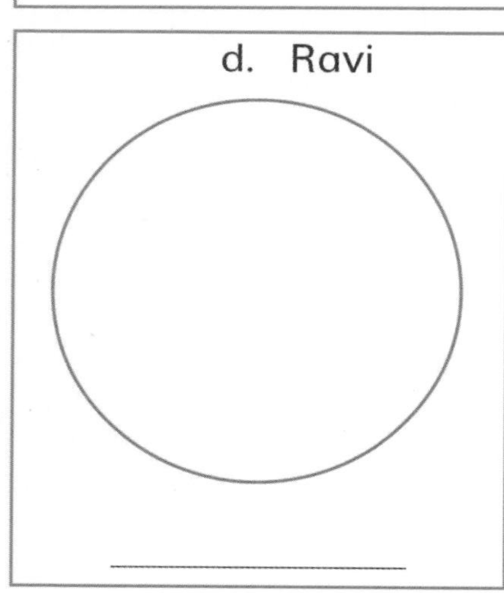

e. Lebo

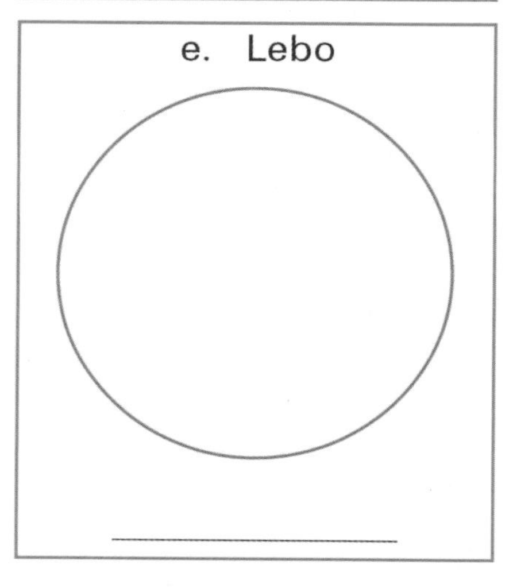

f. Emma

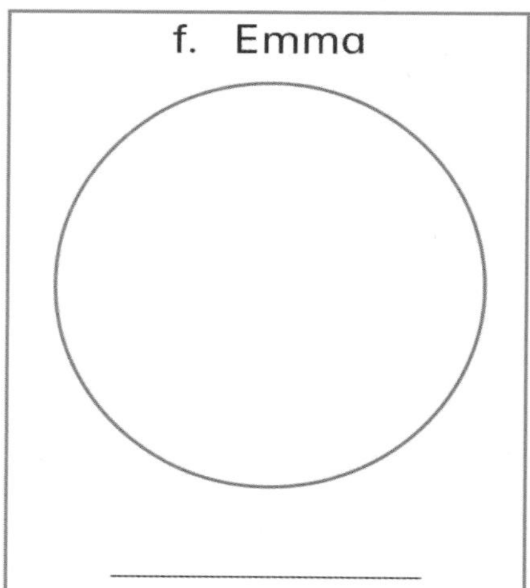

3. Who ate the most pizza? ______________________
   谁吃的披萨最多？
4. Who ate the least pizza? ______________________
   谁吃的披萨最少？

5. There were 12 nuts in each child's bag of nuts. Colour the number of nuts they ate: 每个孩子的坚果袋里有 12 粒坚果。将他们吃的坚果数量涂色：
   a. Jody ate half of her nuts. Jody 吃了一半的坚果。
   b. Mandla ate three quarters of his nuts. Mandla 吃了 $\frac{3}{4}$ 的坚果。
   c. Jaco ate all his nuts. Jaco 吃完了所有的坚果。
   d. Ravi ate a third of his nuts. Ravi 吃了 $\frac{1}{3}$ 的坚果。
   e. Lebo ate two-thirds of her nuts. Lebo 吃了 $\frac{2}{3}$ 的坚果。
   f. Emma only ate a quarter of her nuts. Emma 仅仅吃了 $\frac{1}{4}$ 的坚果。

6. Write the fraction and number of nuts under each bag of nuts.
在每袋坚果的下方，写出分数和坚果的数量。

a. Jody

☐ or ______ nuts

b. Mandla

☐ or ______ nuts

c. Jaco

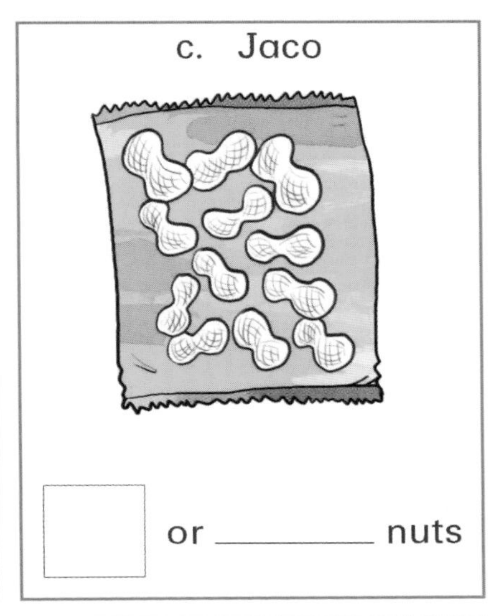

☐ or ______ nuts

d. Ravi

☐ or ______ nuts

e. Lebo

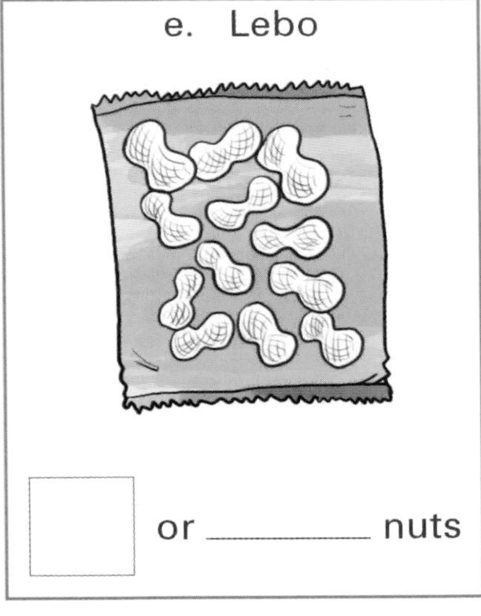

☐ or ______ nuts

f. Emma

☐ or ______ nuts

7. Write these fractions from smallest to biggest on the number line.
在数字线上从小到大写出这些分数。

1, $\frac{1}{2}$, $\frac{1}{4}$, $\frac{3}{4}$

0 ☐ ☐ ☐ ☐

Skill: recognising fractions

Peanuts, please 请，吃花生

These monkeys live at the zoo. They love to eat peanuts. Unfortunately, some of the bags had holes in them so some peanuts fell out.

这群猴子在动物园生活。他们爱吃花生。不幸的是，一些袋子有破洞，因此一些花生掉了出来。

1. Help Ravi work out how many peanuts are left in each bag.
   帮助 Ravi 算出每个袋子里还剩下多少花生。

   a. There were 647 peanuts in this bag. 25 peanuts fell out.
   这个袋子里有 647 粒花生。掉出了 25 粒花生。

   ☐ – ☐ = ☐ peanuts left.
   粒花生。

   b. There were 198 peanuts in this bag. 76 peanuts fell out.
   这个袋子里有 198 粒花生。掉出了 76 粒花生。

   ☐ – ☐ = ☐ peanuts left.
   粒花生。

   c. There were 473 peanuts in this bag. 62 peanuts fell out.
   这个袋子里有 473 粒花生。掉出了 62 粒花生。

   ☐ – ☐ = ☐ peanuts left.
   粒花生。

   d. There were 281 peanuts in this bag. 51 peanuts fell out.
   这个袋子里有 281 粒花生。掉出了 51 粒花生。

   ☐ – ☐ = ☐ peanuts left.
   粒花生。

   e. There were 356 peanuts in this bag. 33 peanuts fell out.
   这个袋子里有 356 粒花生。掉出了 33 粒花生。

   ☐ – ☐ = ☐ peanuts left.
   粒花生。

   f. There were 529 peanuts in this bag. 16 peanuts fell out.
   这个袋子里有 529 粒花生。掉出了 16 粒花生。

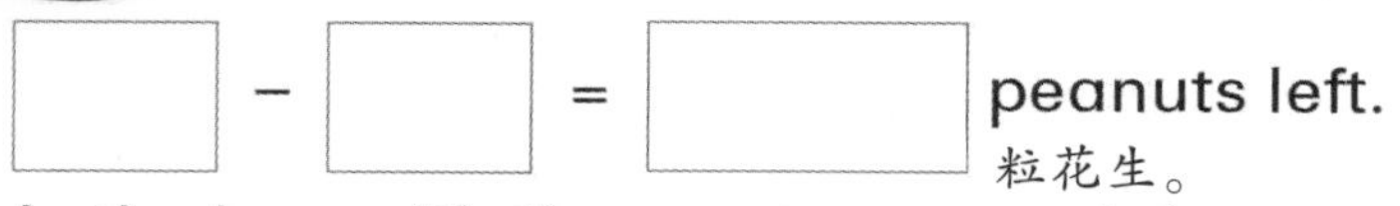

   ☐ – ☐ = ☐ peanuts left.
   粒花生。

2. Circle the bag with the most peanuts left. 把剩余花生最多的那个袋子圈起来。

3. Put a tick next to the bag with the least peanuts left.
   在剩余花生最少的袋子旁边做个记号。

Shape transformers

图形转换

There are shapes everywhere.

无论什么地方都有图形。

Find and label the shapes in the pictures.

在下列图中找出各种图形并标注出来。

找出以下这些图形：三角形、圆形、椭圆形、长方形、六边形和正方形。

Skill: identifying shapes

Time twisters

时间难题

Are you ready for a time quiz? Work out the answers and see whether you are a time whizz!

你准备好做时间测试了吗？算出答案，看看你是否是一个时间神童！

1. Jody went away for the weekend. She left on Friday morning at six o'clock. She arrived back home at noon on Saturday.

   Jody 周末外出度假。她是星期五上午 6 点离开的。她在星期六中午回到家。

   a. How many days was Jody away from home? ☐

   Jody 离开家多少天？

   b. How many hours was she away from home? ☐

   她离开家有多少小时？

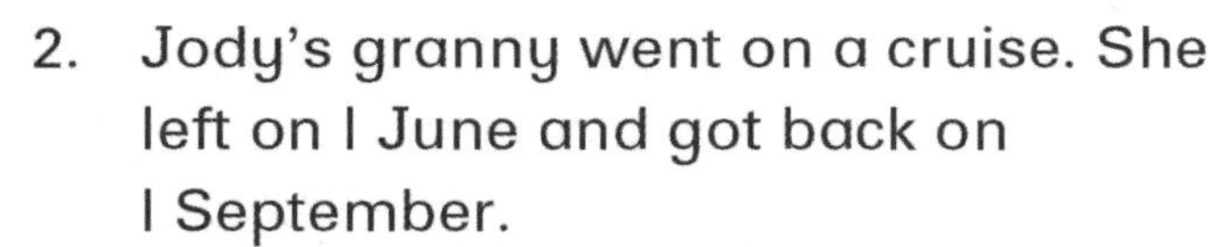

2. Jody's granny went on a cruise. She left on 1 June and got back on 1 September.

   Jody 的奶奶去旅行了，她 6 月 1 日离开，9 月 1 日返回。

   a. How many months was she away? ☐

   她离开了多少月？

   b. How many days was she away? ☐

   她离开了多少天？

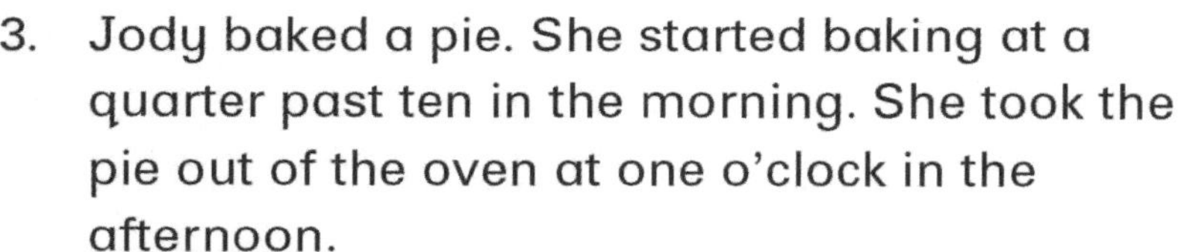

3. Jody baked a pie. She started baking at a quarter past ten in the morning. She took the pie out of the oven at one o'clock in the afternoon.

   Jody 烤了一个馅饼。她在上午 10 点 15 分开始烘烤，下午 1 点钟从烤箱里取出馅饼。

   a. How many minutes did it take to prepare and bake the pie?

   从准备到烤好馅饼共花了多少分钟？

   ☐

   b. Write this in hours and minutes.

   将它用小时和分钟写出来。

_______________ _______________

4. Jody went to the beach by train. The train left at 11:25 a.m. It arrived at the beach at 11:55 a.m.
   Jody 坐火车去了海滩。火车离开的时间是上午 11:25。到达的时间是上午 11:55。
   a. How many minutes did the train trip take? 乘火车花了多少分钟？
   b. How many hours did the train trip take?  乘火车花了多少小时？

5. Jody's school day is from 08:00 until 13:00.
   Jody 上学的时间是从 8:00 到 13:00。
   a. How many hours is she at school each day?
      她每天在学校呆多少小时？
   b. How many hours does she spend at school in a week? 
      她一周在学校呆多少小时？

6. Today is 2 June. Jody's birthday is on 11 August. How many days until Jody's birthday? 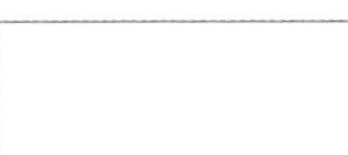
   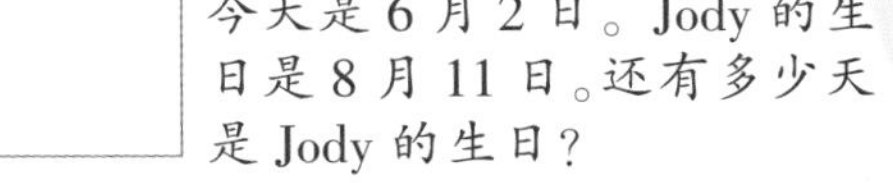
   今天是 6 月 2 日。Jody 的生日是 8 月 11 日。还有多少天是 Jody 的生日？

How many did you get correct?
你做对了多少？

9–11 correct 正确	You are a time whizz! 你是时间神童！
5–8	You are ticking! 你是一个报时钟！
0–4	Try again! 再试着做一遍！

Cut the cake 切蛋糕

This is Jaco's birthday cake.

这是 Jaco 的生日蛋糕。

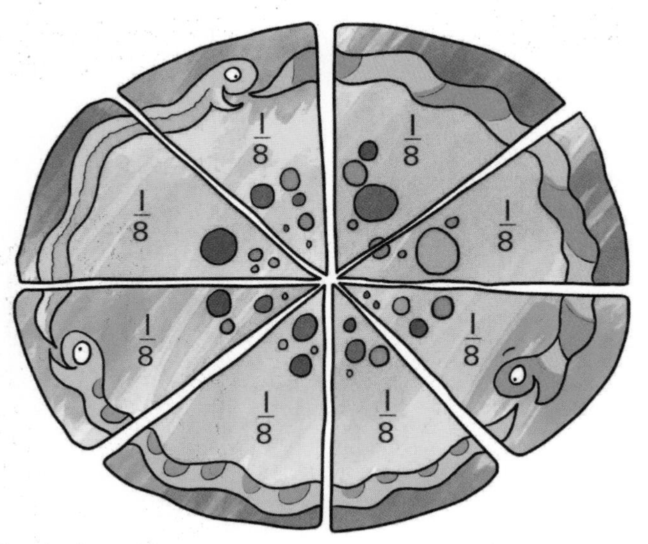

Complete the sentences.

完成这些习题。

1. The cake has been cut into ☐ pieces. 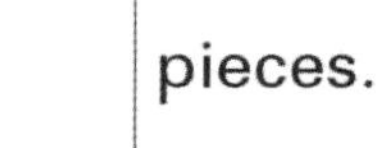
   蛋糕切成了___块。
2. The cake is shared between two people.
   将蛋糕分给 2 个人。

   Each person will get ☐ of the cake.
   每个人分到蛋糕的___。
3. The cake is shared between four people.
   将蛋糕分给 4 个人。

   Each person will get ☐ of the cake.
   每个人分到蛋糕的___。
4. If Jaco eats three slices, he has eaten ☐ of the cake.
   如果 Jaco 吃了 3 块，他吃了蛋糕的___。
5. If he eats two slices, he has eaten ☐ of the cake.
   如果他吃了 2 块，他吃了蛋糕的___。
6. If he eats six slices, he has eaten ☐ of the cake.
   如果他吃了 6 块，他吃了蛋糕的___。
7. If he eats four slices, he has eaten ☐ of the cake.
   如果他吃了 4 块，他吃了蛋糕的___。
8. If he eats $\frac{8}{8}$ of the cake, he has eaten ☐ cake!
   如果他吃了蛋糕的$\frac{8}{8}$，他吃了___蛋糕！

Missing bits 剩下的部分

Term 3

Lebo's teacher had to leave her classroom before she finished her multiplication lesson for the day.

Lebo 的老师在上完乘法课之前,不得不离开教室。

1. Can you be the teacher and finish the work on the board?

   你能成为一名老师并完成黑板上的任务吗?

2. Make five number sentences of your own using these patterns.

   仿照上面的示例,你自己写出 5 道数字习题。

____

____

____

____

____

Skill: multiplication

Smart darts
智能飞镖

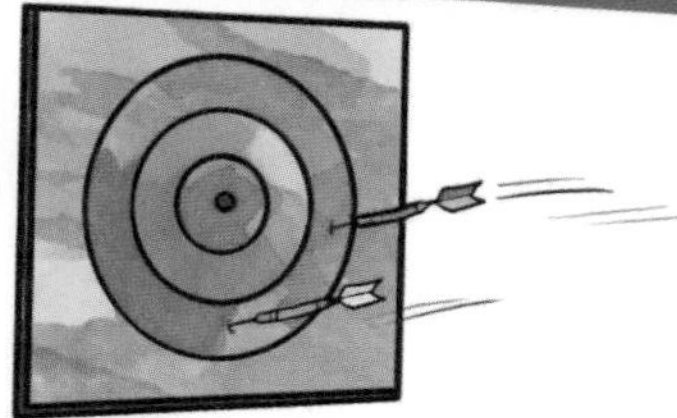

Ravi's dad is playing in a darts competition.
Ravi 的爸爸正在进行飞镖比赛。

1. Which numbers would he need to hit in order to get each of the scores? Write down two different ways to reach each score. You can use the same number more than once. 为了获得每个得分，他需要击中哪些数字？写下两种不同的得分方法，来获得每个得分，你可以重复使用同一个数字。

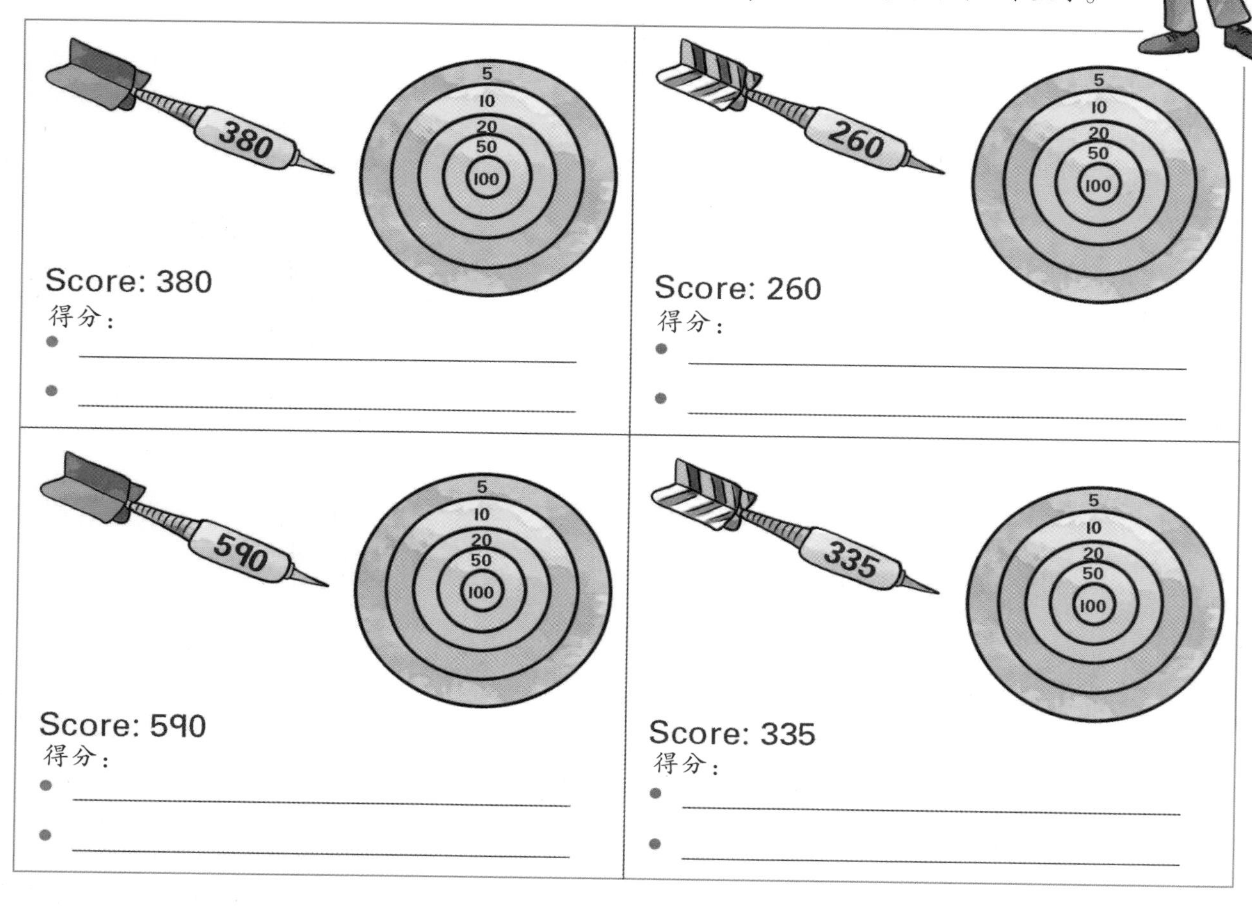

2. What score will Ravi's dad get if his dart hits these numbers? Write the answers on the dart.
如果 Ravi 的爸爸击中了这些数字，那么他将获得多少分？在飞镖上写出答案。

a. $100 + 100 + 100 + 50 + 50 + 10 =$ 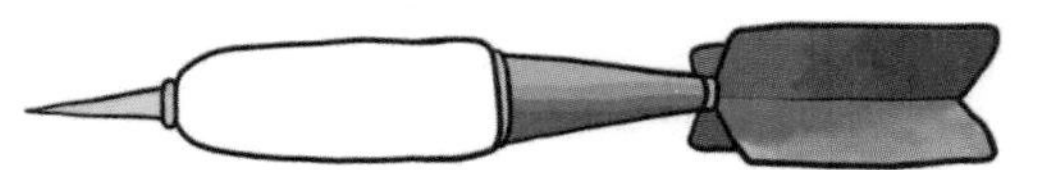

b. $100 + 50 + 10 + 10 + 10 + 10 + 5 =$ 

c. $100 + 100 + 100 + 100 + 100 + 10 + 10 =$ 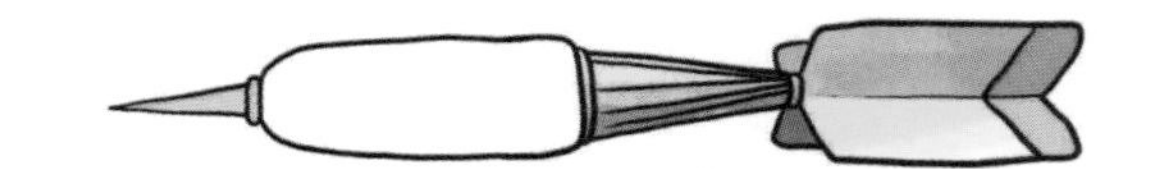

d. $100 + 100 + 50 + 50 + 20 + 20 =$ 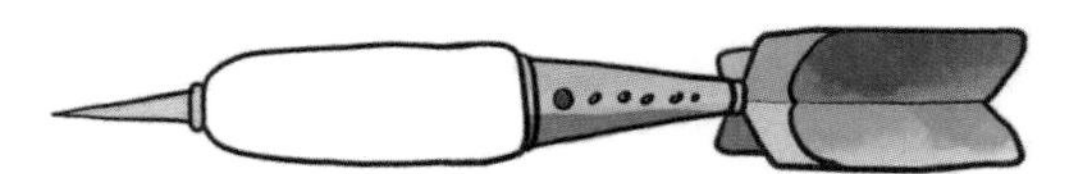

Here are the scores of all the people who took part in the dart competition.
这是飞镖比赛所有参赛者的得分。

Competition scores

比赛得分

139	189	432	651
452	456	145	345
291	165	352	543
446	267	412	
632	672	430	

3. Now write them in order from highest to lowest.
现在按照从高到低的顺序写出它们。

Winners Board

获奖者告示板

1st ______ 2nd ______

3rd ______ 4th ______

5th ______ 6th ______

7th ______ 8th ______

9th ______ 10th ______

11th ______ 12th ______

13th ______ 14th ______

15th ______ 16th ______

17th ______ 18th ______

4. What is the highest score? ☐
分数最高的是？

5. What is the lowest score? ☐
分数最低的是？

Skills: repeated addition, ordering numbers

Map a bedroom

绘制卧室平面图

Princess Lebo's favourite colour is purple. She would like to decorate her bedroom using a lot of purple!

Lebo 公主最喜欢的颜色是紫色。她想用许多紫色来装饰她的卧室!

1. Look at the key for the map.
   看一下平面图的图例。
2. Draw one or more of each of these things to make a map of Princess Lebo's bedroom.
   在 Lebo 公主的卧室内画出每件物品,一件或多件均可,制作一幅平面图。

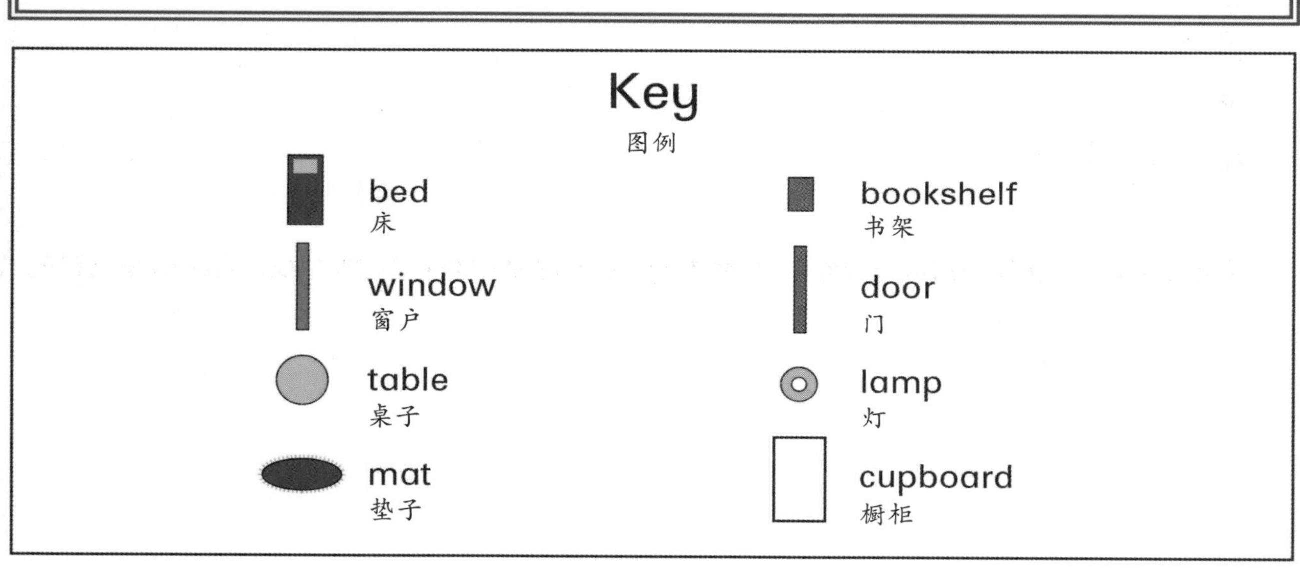

3. Now create a map of your own bedroom. 现在画出你自己卧室的平面图。
Don't forget to design a key for your map. 不要忘了给你的平面图设计图例。

Key
图例

Building blunders

建筑上的错误

Geraldine Giraffe built a new house but she is having some building problems.

长颈鹿 Geraldine 盖了一个新房子，但是她有一些建筑上的问题。

1. Write down the height measurements for Geraldine's house.
   写出 Geraldine 的房子的高度测量值。
2. As you can see, the house is too small for Geraldine. She asked the builder to build it again. She said she wanted a house that is double the size. Double the numbers so that the builder has the correct measurements.
   正如你看到的，这个房子对于 Geraldine 来说太小了。她让建筑师重新建造。她说，她希望新建的房子比原来的大一倍。把这些数字加倍，建筑师便有了正确的测量值。

Height of Geraldine's house
Geraldine 房子的高度

Roof ______________ cm
屋顶

Wall ______________ cm
墙

Door ______________ cm
门

Garage ______________ cm
车库

Height of new house 新房子的高度

Roof ______________ cm
屋顶

Wall ______________ cm
墙

Door ______________ cm
门

Garage ______________ cm
车库

3. Geraldine's pet tortoise, Terry, asked her to build him a little house to play in. She decided to use the left-over wood to make a house half the size of her old house for Terry. Can you halve the numbers so that she has the correct measurements?

Geraldine 的宠物龟 Terry，请她给他盖一间小房子在里面玩儿。她决定用剩余的木料为 Terry 建一间比她原来房子小一半的房子。你能把这些数字减半使她有正确的测量值吗？

Height of Terry's house

Terry 房子的高度

Roof ________________ cm
屋顶

Wall ________________ cm
墙

Door ________________ cm
门

Garage ________________ cm
车库

4. Now Geraldine needs to paint her new house and Terry's house. She knows how much paint she needed for her old house. Help her work out how much paint she will need now.

现在 Geraldine 需要粉刷她的新房子和 Terry 的房子。她知道她的老房子需要用多少涂料。帮她计算现在需要多少涂料。

	Old house 老房子	New house 新房子	Terry's house Terry 的房子
Roof 屋顶	346 litres 升		
Wall 墙	484 litres 升		
Door 门	150 litres 升		
Garage 车库	264 litres 升		

Skill: doubling and halving numbers

It's a six 有几个 6

The jungle creatures held their annual cricket tournament. There was great excitement as many runs were scored.

丛林里的动物们举行了他们每年一次的板球比赛。许多项竞赛都能得分,特别让人兴奋。

1. Work out how many 2s, 4s or 6s were scored by each team.
   计算出每个队的得分中有多少个 2,多少个 4 或者多少个 6。

All the Raccoons' runs were 6s.
浣熊的得分是 6 的倍数。
How many 6s did the team score?
这个队的得分有多少个 6?

All the Chimps' runs were 6s.
黑猩猩的得分是 6 的倍数。
How many 6s did the team score?
这个队的得分有多少个 6?

All the Squirrels' runs were 4s.
松鼠的得分是 4 的倍数。
How many 4s did the team score?
这个队的得分有多少个 4?

All the Frogs' runs were 2s.
青蛙的得分是 2 的倍数。
How many 2s did the team score?
这个队的得分有多少个 2?

2. Write the team names in order according to the number of runs they scored. Start with the team that scored the most.
   按照他们得分多少的顺序写下队名。从得分最多的队开始。

   a. ______________________________

   b. ______________________________

   c. ______________________________

   d. ______________________________

Around the sides

环绕周边

When you measure around the outside of an object you are measuring its perimeter.
当你环绕物体外边测量时，你就是在测量它的周长。

Which one of these interesting shapes do you think will have the longest perimeter? 这些有趣的图形中，你认为周长最长的是哪一个？

Find: string, a ruler. 找出：绳子，格尺。

1. Use a different piece of string to measure the perimeter of each shape.
   用不同的绳子去测量每个图形的周长。
2. Now measure the string against a ruler to find out how many centimetres each perimeter is.
   现在对着格尺来测量绳子，找出每个周长是多少厘米。

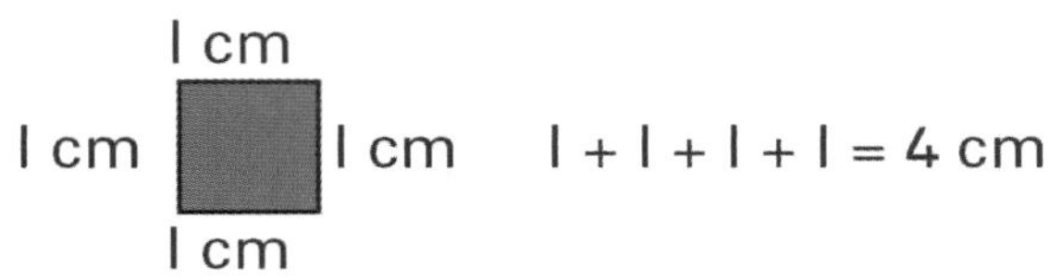

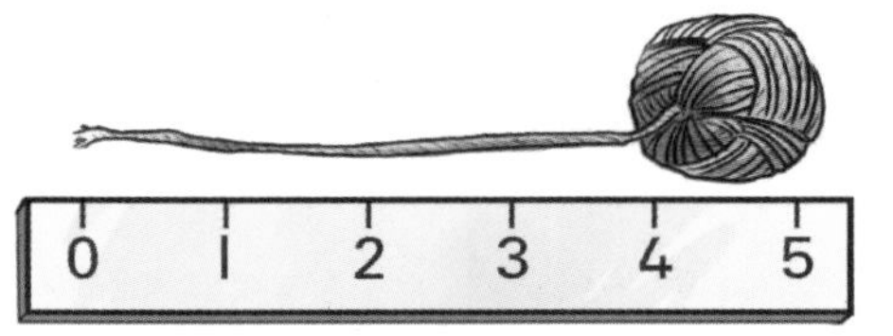

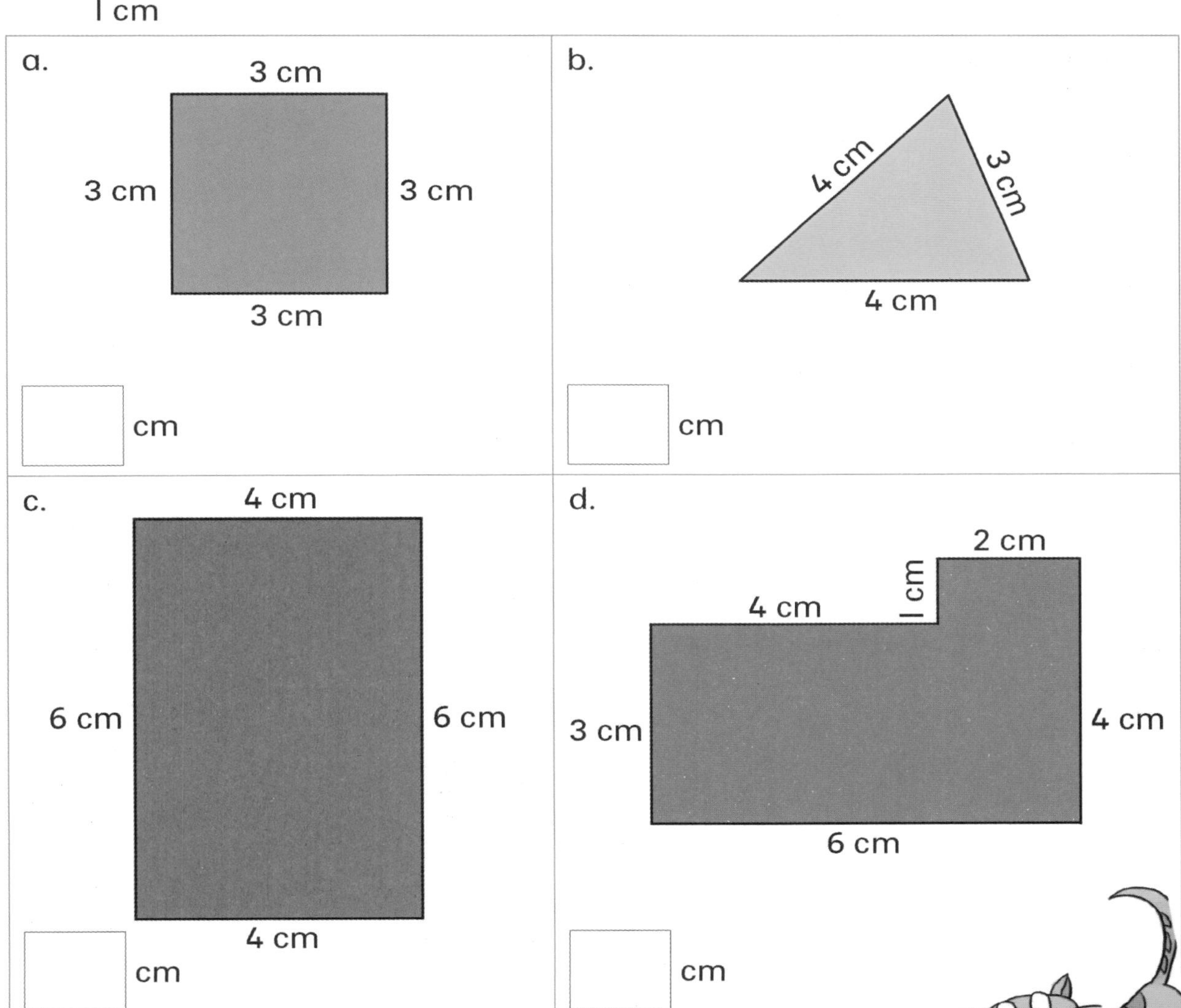

3. Put a tick next to the shapes with the longest perimeters.
   在周长最长的图形旁边画一个记号。

Term 3

Skill: measuring perimeter

Shapes around us

我们周围的图形

There are many shapes in the things around us.
在我们周围的生活中有许多图形。

What shapes and objects can you see in these pictures?
在这些图中你能看到什么图形和物体？

1.

Shapes and objects
图形和物体

circle 圆形	pentagon 五边形
triangle 三角形	diamond 菱形
square 正方形	pyramid 锥体
cube 立方体	cylinder 圆柱体
rectangle 长方形	sphere 球体

2.

3.

4.

5.

Skill: identifying shapes

Snap 捉对儿

Term 4

Do you like to play Snap? Mandla has a set of Snap cards but one card does not fit in. Which one is it?

你喜欢玩捉对儿吗？Mandla 有一副捉对儿卡片，但有一张卡片不配对儿。它是哪张？

1. Find the Snap pairs by completing the sums.
   完成算术题并找出一对儿的卡片。
2. Colour each pair of cards the same colour.
   把每一对儿卡片涂成相同的颜色。

3. Circle the card that does not fit in.
   圈出不配对儿的卡片。

Skill: multiplication

Ready, aim, fire! 准备，瞄准，开火！

Watch out for the cannon balls!
小心，炮弹！

1. Find the cannon balls that belong to each cannon.
   Draw a line from the cannon balls to the cannon.
   找出每门大炮发出的炮弹。在大炮和炮弹之间画一条线。

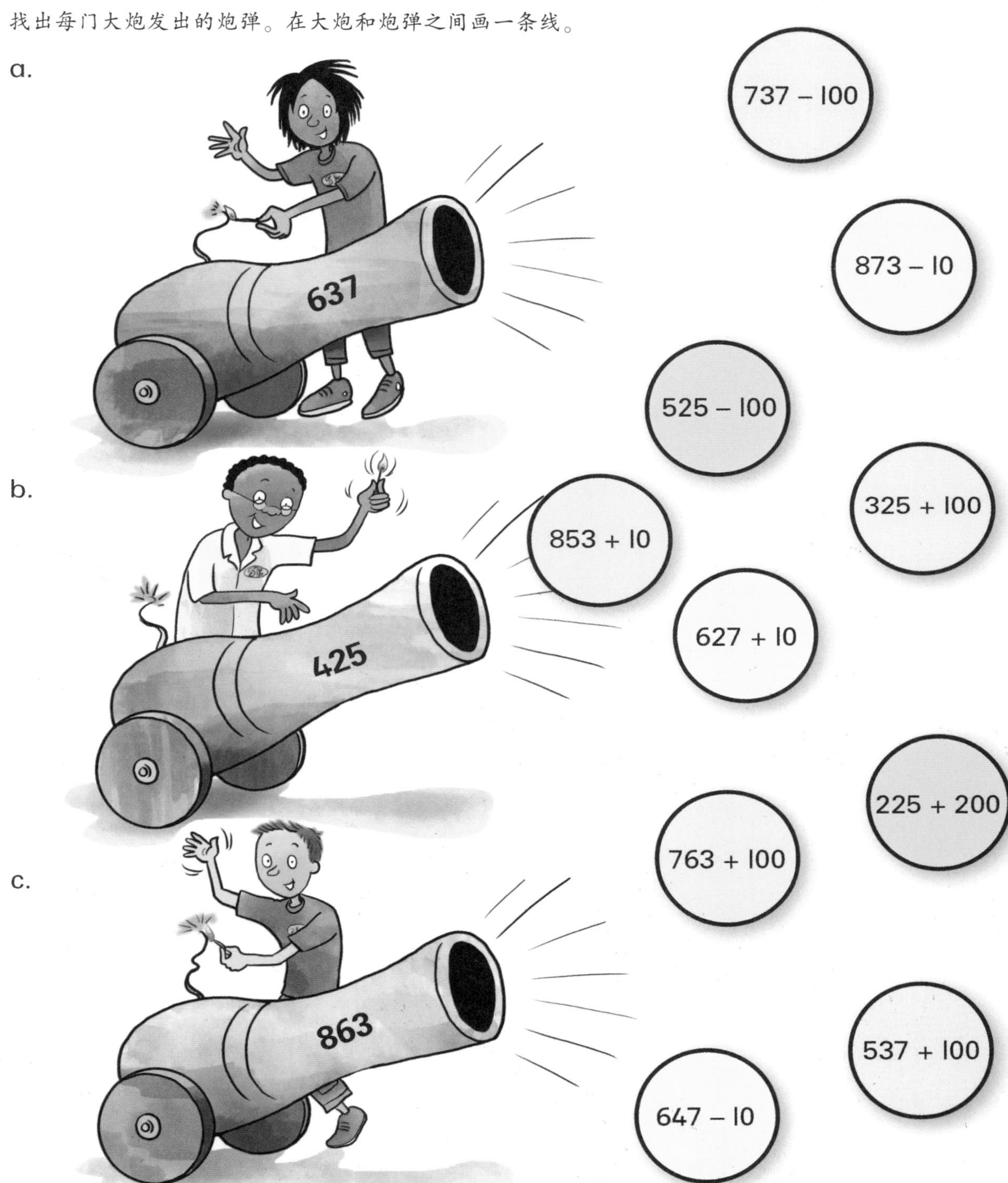

2. Which cannon shot the most balls? ______________________
   哪门大炮射出的炮弹最多？

Skill: addition and subtraction

Crow's nest 瞭望台

Can you see the crow's nest?

你能看见瞭望台吗？

1. Help each pirate reach the lookout point by counting in his favourite way. As you count, write the numbers in the space on the net until you reach the nest.
   请帮助每个海盗用他们喜欢的计数方式到达瞭望台。你在计数时就把数字写在网格内，直至到达瞭望台。

I like to count in 3s.

我喜欢用 3 的倍数计数。

3

I can only count in 4s.

4

我只会用 4 的计数倍数。

100

我喜欢用 100 的倍数计数。

I like counting in 100s.

I love counting in 25s.

25

我喜欢用 25 的倍数计数。

2. Polly, the captain's parrot, has seen the pirates coming. She is trying to escape from the crow's nest. Polly likes to count backwards in 3s. Help her climb down the mast safely.

Polly,是船长的鹦鹉,看到海盗来了。她试图逃离瞭望台。Polly 喜欢按 3 的倍数倒着计数。帮她安全地爬下桅杆。

3. This pirate is still learning to count in 1 000s. He keeps getting stuck. Can you help him?

这个海盗仍在学习按 1000 的倍数计数。他常被卡住数不清。你能帮助他吗?

1 002, 2 002, ________, ________,

________, ________, ________,

________, ________.

1 061, 2 061, ________, ________,

________, ________, ________,

________, ________.

Skill: counting

Number plate maths

号码牌数学

Have you ever added the numbers on car number plates?

你曾经把车牌号上的数字相加过吗？

1. Add the black number and the blue number on each number plate. Write the sum and the answer below.
   将每个号码牌上的黑色数字与蓝色数字相加。写出算术题和答案。
2. Which number is the biggest? Colour the answer block green.
   哪个数最大？将答案格涂成绿色。
3. Which number is the smallest? Colour the answer block orange.
   哪个数最小？将答案格涂成橙色。

SP 2463

246 + 3 = ☐

DV 1671

☐ + ☐ = ☐

LX 7526

☐ + ☐ = ☐

RT 6412

☐ + ☐ = ☐

VT 3407

☐ + ☐ = ☐

CR 9312

☐ + ☐ = ☐

GT 8724

☐ + ☐ = ☐

RW 9431

☐ + ☐ = ☐

Sinking ships 沉船

Stop the ships from sinking. Get rid of some crates to make the ships lighter. 阻止船的下沉。从船上放下一些箱子让船轻一些。

1. Make the ships lighter. Write the answers.
   把船减轻一些。写出答案。

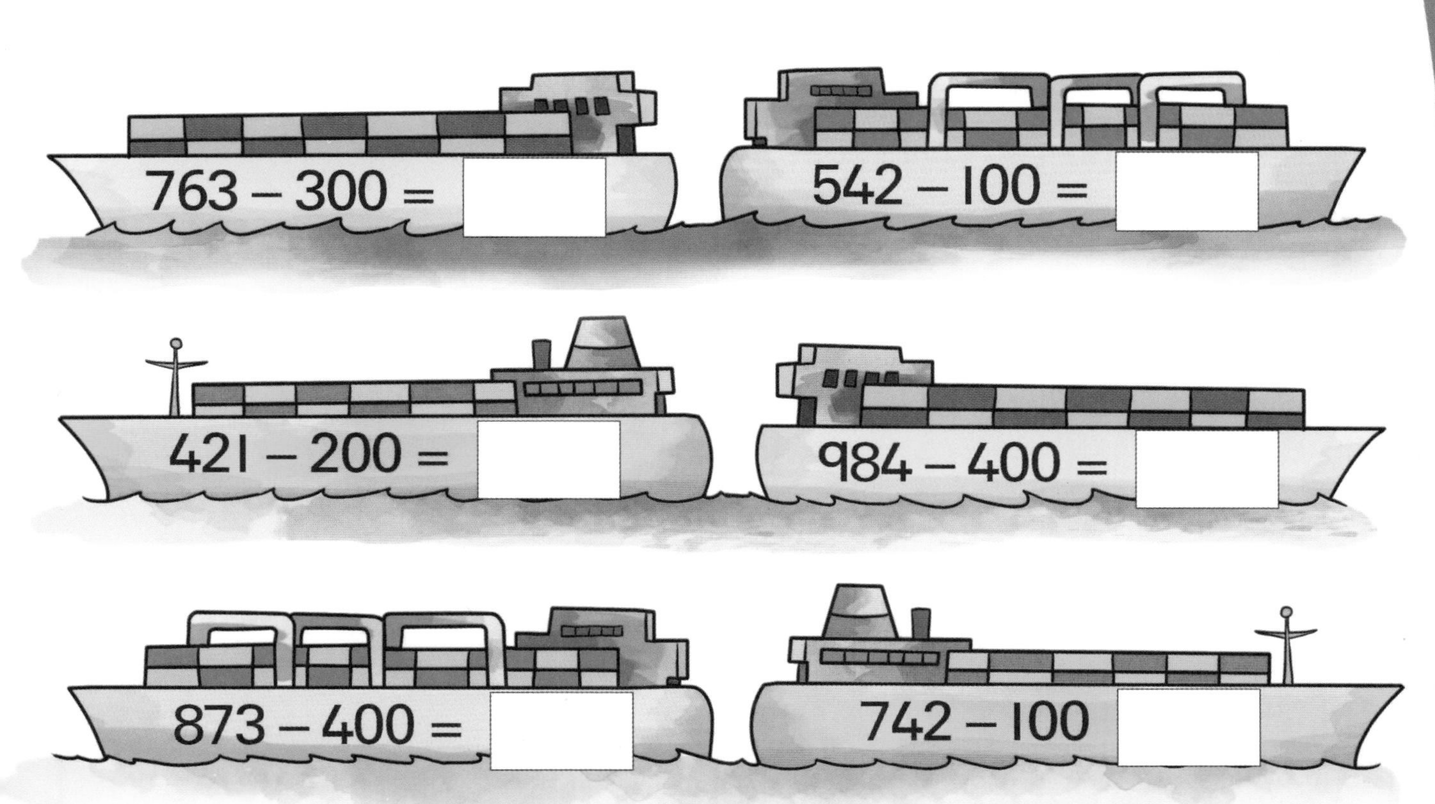

2. Any ships carrying more than 500 crates will sink.
   承载有 500 个以上箱子的任何只船都会下沉。
   a. How many ships will sink?
      有多少只船会下沉？
   b. How many ships will be safe?
      有多少只船是安全的？

Term 4

Skill: subtraction

A fraction of time

时间分数

Fractions of time are easy to see on a clock.
时间分数在时钟上轻易可见。

1. Work out the fractions. 计算出下列分数。

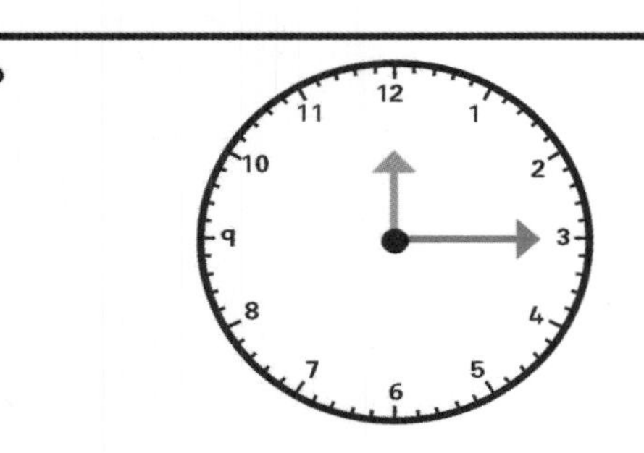

$\frac{1}{4}$ hour 小时 = [ ] minutes 分钟

Colour in **a quarter** of an hour.
将$\frac{1}{4}$小时那部分涂色。

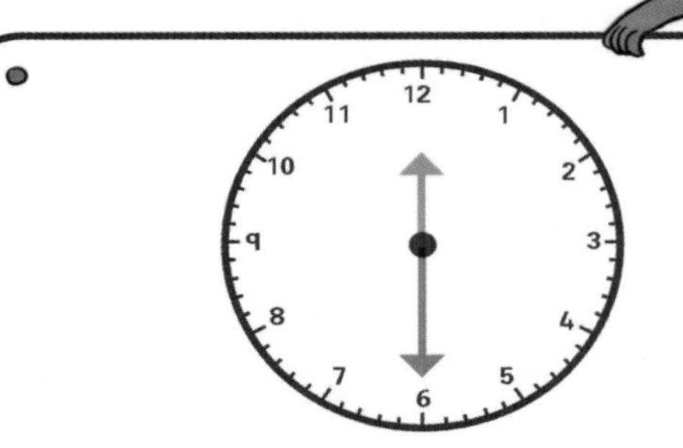

$\frac{1}{2}$ hour 小时 = [ ] minutes 分钟

Colour in **half** of an hour.
将半小时那部分涂色。

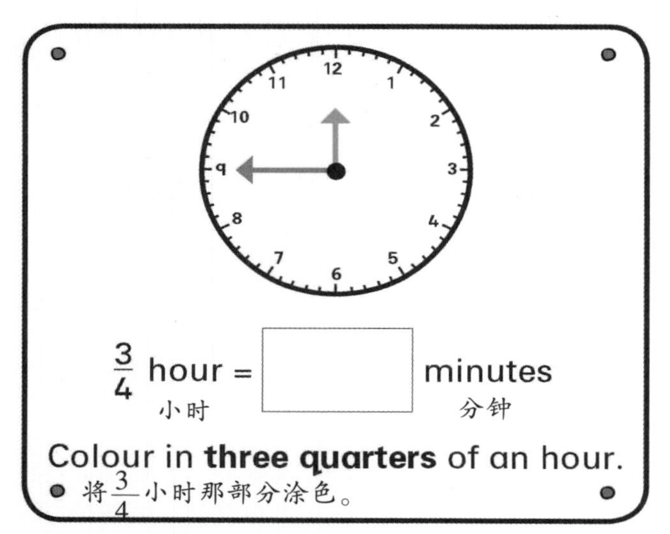

$\frac{3}{4}$ hour 小时 = [ ] minutes 分钟

Colour in **three quarters** of an hour.
将$\frac{3}{4}$小时那部分涂色。

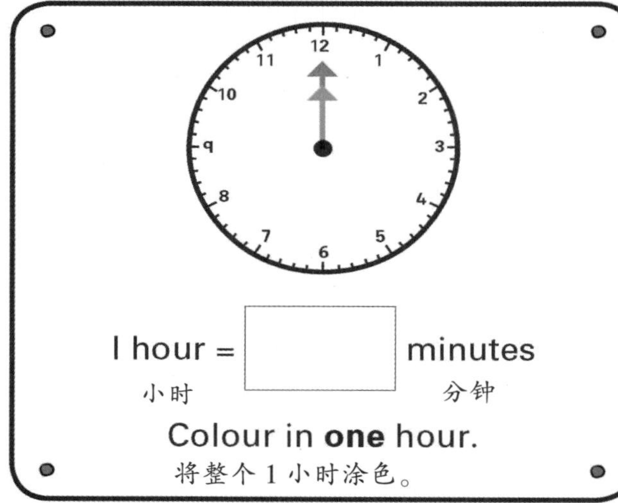

1 hour 小时 = [ ] minutes 分钟

Colour in **one** hour.
将整个1小时涂色。

2. Complete the sentences. Use these words:
完成下列计算。用这些词：

half $\frac{1}{2}$	one 1	three quarters $\frac{3}{4}$	quarter $\frac{1}{4}$

a. 12:00 to 12:30 = ______________________________ an hour.

b. 12:00 to 12:45 = ______________________________ of an hour.

c. 12:00 to 13:00 = ______________________________ hour.

d. 12:00 to 12:15 = ______________________________ of an hour.

A slice of time 一段时间

The things we do each day take up different amounts of time. Can you work out what fraction of time these activities take?
我们每天做的事情占用了不同的时间。你能计算出这些活动占用的时间分数吗？

1. It takes Jody forty-five minutes to wash the car. What fraction of an hour is this?
Jody 洗车花费了 45 分钟。这是 1 小时的几分之几？

 of an hour
小时

Draw the hands on the clock and colour in the fraction.
在时钟上画出指针，并给这部分涂色。

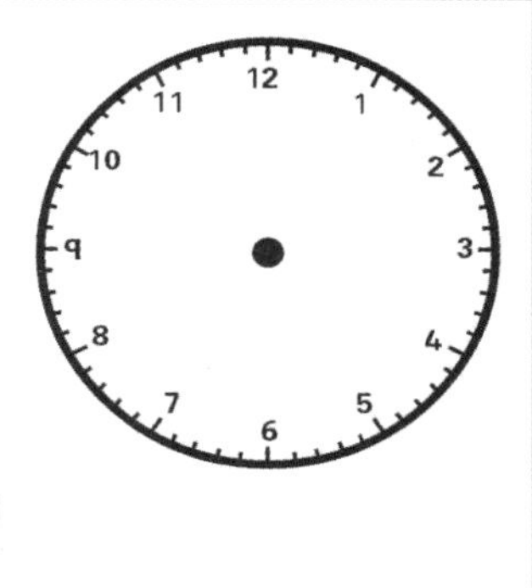

2. It takes Ravi a quarter of an hour to eat breakfast. How many minutes is that?
Ravi 吃早餐花费了 $\frac{1}{4}$ 小时。那是多少分钟？

 minutes
分钟

Draw the hands on the clock and colour in the fraction.
在时钟上画出指针，并给这部分涂色。

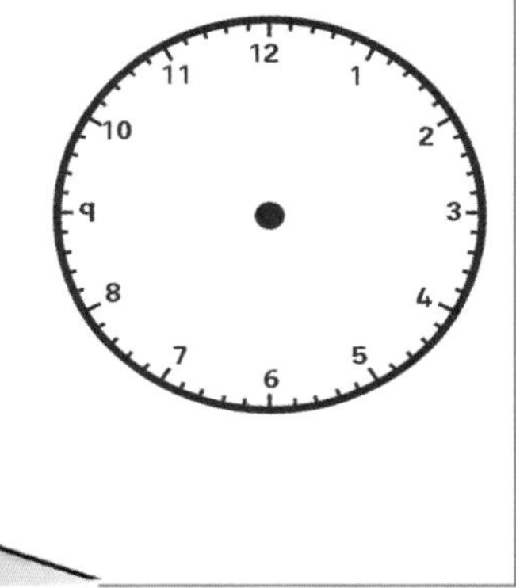

3. Mandla had a nap from quarter past three to quarter past four. How long did he sleep?
Mandla 从 3 点 15 分到 4 点 15 分小睡了一会儿。他睡了多久？

 minutes or 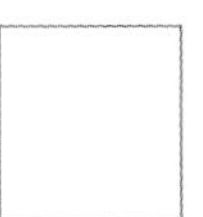 hour
分钟 小时

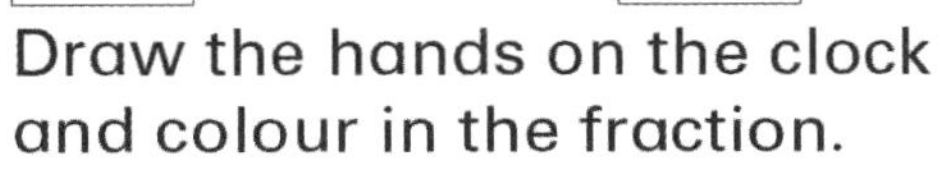
Draw the hands on the clock and colour in the fraction.
在时钟上画出指针，并给这部分涂色。

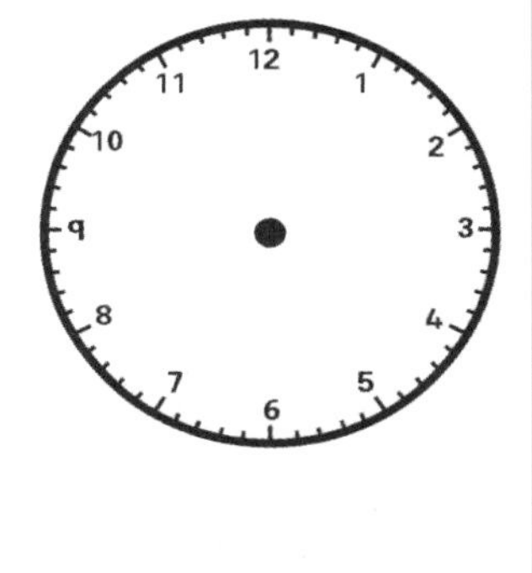

4. Lebo read her book from quarter to nine to quarter past nine. How long did she read?
Lebo 从 8 点 45 分到 9 点 15 分在读书。她读了多久？

 minutes or  an hour
分钟 小时

Draw the hands on the clock and colour in the fraction.
在时钟上画出指针，并给这部分涂色。

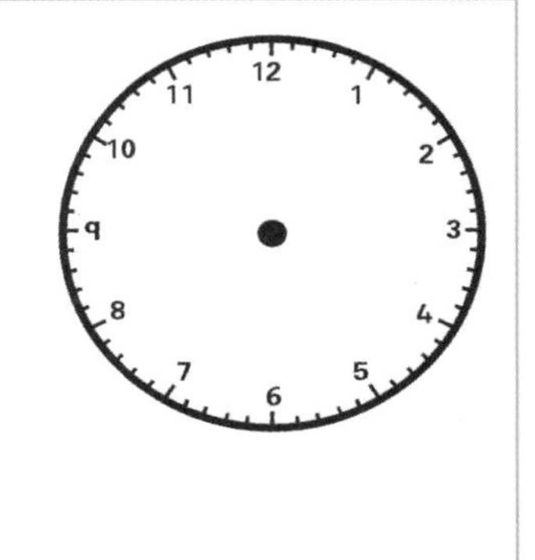

Pyramid puzzles

金字塔谜题

Building pyramids is very hard work!

建造金字塔是非常艰苦的工作！

Help build these pyramids. Use different numbers that together make the number at the top of each pyramid.

请帮助建造这些金字塔。每行用不同的数字相加，使其等于金字塔顶端的数字。

Each row of blocks must add up to the number at the top of the pyramid.

每行砖块上的数字相加必须等于金字塔顶端的数字。

Skill: decomposing numbers

All bottled up 装满瓶子

You can learn about millilitres and litres while having water fun!
在玩儿水的游戏中,你可以了解毫升和升。

Find these empty cups and bottles: 找出这些空的杯子和瓶子:

l litre bottle
1L 瓶子

750 millilitre bottle
750mL 瓶子

500 millilitre bottle
500mL 瓶子

l25 millilitre cup
125mL 杯子

l00 millilitre cup
100mL 杯子

1. Estimate. First write down your guess for each question.
估计值。首先写出每道题的猜想。
2. Then use the cup to fill the bottle.
然后用杯子装满瓶子。
3. Measure and write down the measurement.
测量并写下测量值。

a. How many l00 mL cups fill a 500 mL bottle?
多少个 100mL 的杯子能装满 500mL 的瓶子?

Estimate 估计值 ☐ Measure 测量值 ☐

b. How many l00 mL cups fill a l L bottle?
多少个 100mL 的杯子能装满 1L 的瓶子?

Estimate 估计值 ☐ Measure 测量值 ☐

c. How many l25 mL cups fill a 750 mL bottle?
多少个 125mL 的杯子能装满 750mL 的瓶子?

Estimate 估计值 ☐ Measure 测量值 ☐

d. How many l25 mL cups fill a 500 mL bottle?
多少个 125mL 的杯子能装满 500mL 的瓶子?

Estimate 估计值 ☐ Measure 测量值 ☐

e. How many l25 mL cups fill a l L bottle?
多少个 125mL 的杯子能装满 1L 的瓶子?

Estimate 估计值 ☐ Measure 测量值 ☐

Skill: measuring capacity

Shereen's shoe shop

Shereen 的鞋店

Shereen sells many beautiful shoes in her shoe shop. She has shoes for spiders, bees, tigers and people.

Shereen 的鞋店出售许多漂亮的鞋。她出售蜘蛛、蜜蜂、老虎和人穿的鞋。

Help her work out how many shoes she has for each group.

帮她算一算每组有多少只鞋。

38 只蜜蜂的靴子

29 Humans' high heels

29 只女人的高跟鞋

Complete the sentences.

完成下列句子。

1. I spider needs ____ sandals.
   1 只蜘蛛需要 ____ 只凉鞋。

2. I bee needs ____ boots.
   1 只蜜蜂需要 ____ 只靴子。

3. I tiger needs ____ takkies.
   1 只老虎需要 ____ 只大头鞋。

4. I human needs ____ high heels.
   1 个人需要 ____ 只高跟鞋。

5. Shereen has enough sandals for ____ spiders.
   Shereen 的凉鞋可供 ____ 只蜘蛛。

   She will have ____ sandals left over.
   她将剩下 ____ 只凉鞋。

6. Shereen has enough boots for ____ bees.
   Shereen 的靴子可供 ____ 只蜜蜂。

   She will have ____ boots left over.
   她将剩下 ____ 只靴子。

7. Shereen has enough takkies for ____ tigers.
   Shereen 的大头鞋可供 ____ 只老虎。

   She will have ____ takkies left over.
   她将剩下 ____ 只大头鞋。

8. Shereen has enough high heels for ____ humans.
   Shereen 的高跟鞋可供 ____ 个女人。

   She will have ____ high heels left over.
   她将剩下 ____ 只高跟鞋。

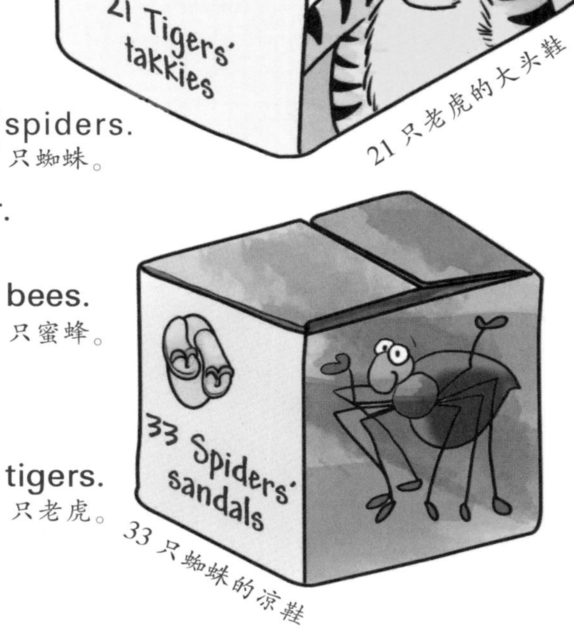

21 只老虎的大头鞋

33 只蜘蛛的凉鞋

Skill: division with remainders

Going batty 狂飞的蝙蝠

Bats love to hide in dark places. 蝙蝠喜欢隐藏在黑暗中。
Can you see them flying around in the cave? 你能看清在山洞中到处飞行的蝙蝠吗?

1. Fill in the missing numbers.
   填上缺少的数字。

2. It is night time so it is safe for the bats to come out of the cave. Write the answers so that each one can leave the cave.
对于蝙蝠来说，晚上从山洞出来是安全的。
写出答案，以便每个蝙蝠都可以离开山洞。

Skill: multiplication

Smooth 'n cool 又光滑又凉爽

You are going to have a stall at your school fair. You have decided to sell ice-cream, milkshake and yoghurt. You will have three flavours: chocolate, vanilla and strawberry. Work out how many of each item you will need!

你打算在学校市场摆一个货摊。你决定卖冰淇淋、奶昔和酸奶。你有 3 种口味：巧克力、香草和草莓。算出每个商品你需要多少！

1. Ask your family and friends to choose one item and one flavour. Write down the combination they choose.
   让你的家人和朋友选择一种商品和一种口味。写下他们选择的组合。

	Name 名称	Item 商品	Flavour 口味
1			
2			
3			
4			
5			
6			

2. How many of each flavour were chosen? 每一种口味选择了多少？

chocolate 巧克力 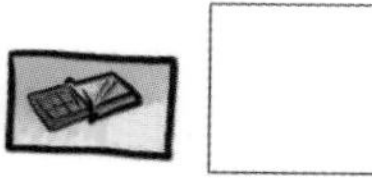 ☐ vanilla 香草 

☐ strawberry 草莓  

3. How many of each item were chosen? 每一种商品选择了多少？

ice-cream 冰淇淋  ☐ milkshake 奶昔  ☐ yoghurt 酸奶  ☐

4. Use the information you collected to complete the graph. Colour in one block for each person's choice.
用你收集的信息完成这个表格。将每个人选择的方框涂色。

6									
5									
4									
3									
2									
1									
	chocolate 巧克力	vanilla 香草	strawberry 草莓	chocolate 巧克力	vanilla 香草	strawberry 草莓	chocolate 巧克力	vanilla 香草	strawberry 草莓

5. Which combination was the most popular? ______________________
哪个组合最受欢迎?

6. Which combination was least popular? ______________________
哪个组合最不受欢迎?

Skill: working with graphs

Wanda, the witch 女巫 Wanda

Wanda, the witch, bought too many ingredients for her spell. Look at what she bought.

女巫 Wanda，为了迷惑人买了许多用品。看看她买了些什么。

Wanda's shopping list

Wanda 的购物清单

48 bats 48 只蝙蝠

18 snakes 18 条蛇

36 bottles of potion
36 瓶药水

24 bowls of swamp mud
24 碗泥浆

30 hairs 30 顶假发

90 newts' ears 90 只蝾螈耳朵

66 magic drops 66 滴魔法水

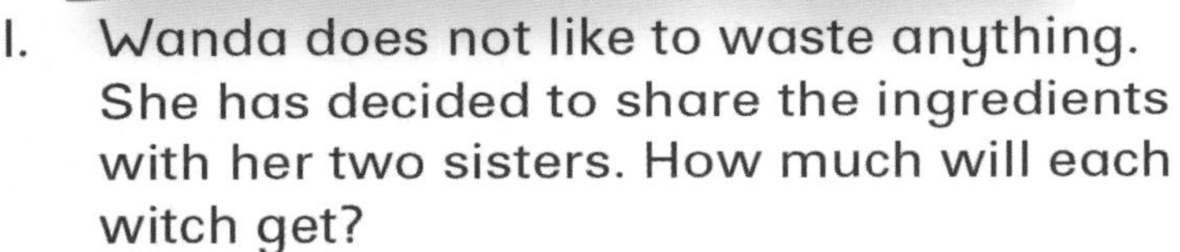

1. Wanda does not like to waste anything. She has decided to share the ingredients with her two sisters. How much will each witch get?

   Wanda 不喜欢浪费东西。她决定将这些用品分给她的两个姐妹。每个女巫得到多少？

   Each of the witches will get:

   每个女巫将会得到：

☐ bats
只蝙蝠

☐ snakes
条蛇

☐ bottles of potion
瓶药水

☐ bowls of swamp mud
碗泥浆

☐ hairs
顶假发

☐ newts' ears
只蝾螈耳朵

☐ magic drops
滴魔法水

2. Three of Wanda's friends arrive for a visit. Share the same ingredients between all the witches. Wanda 的 3 个朋友来看她。将这些用品平分给所有的女巫。

Each of the witches will get: 每个女巫将得到：

☐	bats 只蝙蝠
☐	snakes 条蛇
☐	bottles of potion 瓶药水
☐	bowls of swamp mud 碗泥浆
☐	hairs 顶假发
☐	newts' ears 只蝾螈耳朵
☐	magic drops 滴魔法水

3. Wanda really wishes that she just had to share the ingredients with her best friend, Wilfred the wizard. How much would they each get if just Wanda and Wilfred shared the ingredients?

Wanda 真的希望只和她最好的朋友——男巫 Wilfred 分享这些物品。如果只是 Wanda 和 Wilfred 分享这些物品，他们每个人能得到多少？

Each friend will get: 每个朋友将会得到：

☐	bats 只蝙蝠
☐	snakes 条蛇
☐	bottles of potion 瓶药水
☐	bowls of swamp mud 碗泥浆
☐	hairs 顶假发
☐	newts' ears 只蝾螈耳朵
☐	magic drops 滴魔法水

Awesome area
了不起的面积

1. Colour this square green:
   将这个正方形涂成绿色：

You have coloured the area of the square!
The space inside the perimeter of a shape is called the area.
你已经在这个正方形内涂色！图形周长里面的区域叫做面积。

Let's measure some more area. 让我们再多测几个面积。

2. Measure the area of the floor of a room. Use sheets of newspaper that are all the same size. Lay them down like this:
   测量一下房子地板的面积。像下图所示这样平铺几张大小相同的报纸：

The area of the floor is ☐ sheets of newspaper.
地板的面积是 ☐ 张报纸。

3. Now measure the area of a table. Use sheets of A4 paper.
   现在来测量一下这张桌子的面积。用几张 A4 纸来测量。

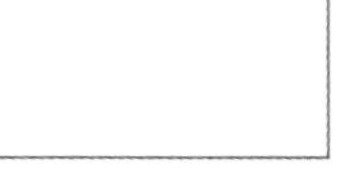

The area of the table is ☐ sheets of paper.
桌子的面积是 ☐ 张纸。

Mosaic mat 马赛克地垫

A mosaic is a pattern usually made up of different coloured tiles packed closely together.

马赛克通常是一种由不同颜色瓷砖贴在一起的图案。

Follow the steps to make your own mosaic mat!

按下面的步骤制作出你自己的马赛克地垫！

1. Cut out ten different-coloured paper tiles that look like this one.
   照下图所示剪出 10 块不同颜色的纸板砖。

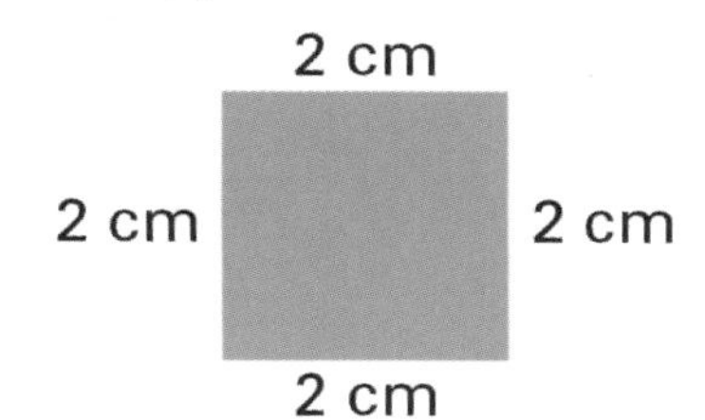

2. Use another piece of paper to make the bottom of your mat. Make it this size:
   用另一张纸当做你地垫的底部。做成这样大小：

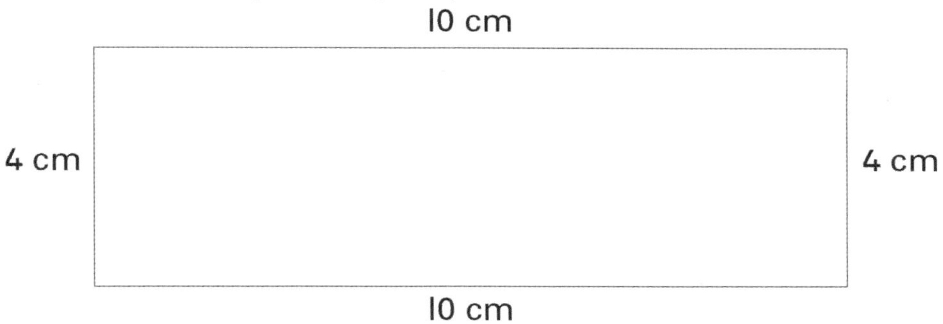

3. Place your tiles on your mat. Make sure all of them fit.
   将你的纸板砖放在地垫上。一定要把所有纸板砖对合好。

4. Glue down the tiles.
   用胶水贴牢纸板砖。
5. Complete the sentence:
   完成这个句子：
   The area of my mat is  tiles.
   我的这块地垫用了　　块纸板砖。

Summer fashion fun

夏季时装游戏

Lebo wanted to know what her friends like to wear most on warm summer days. She asked them to choose from:

Lebo 想知道她的朋友们在温暖的夏日最喜欢穿什么服装。她让他们从下面图片中做出选择：

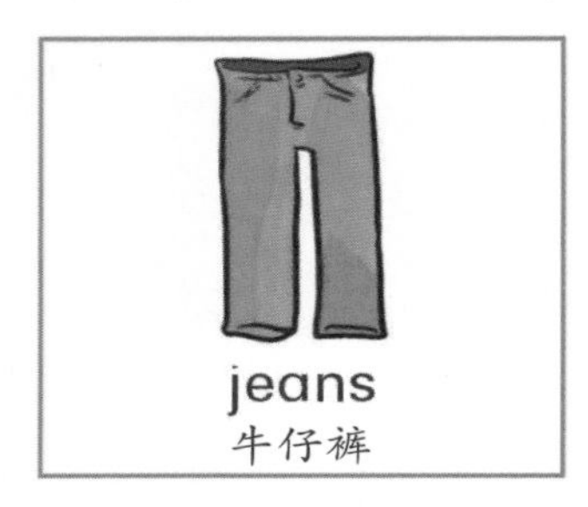

jeans
牛仔裤

T-shirt
T 恤

shorts
短裤

tank top
背心

Lebo wrote down what her friends chose.
Lebo 写下了她朋友们的选择。

1. Add up the totals. 合计如下。

Name 名字	Jeans 牛仔裤	Shorts 短裤	T-shirt T 恤	Tank top 背心
Lesedi		X		X
Emma	X		X	
Bailey	X			X
Teresa	X			X
Poovan		X		X
Nadis	X		X	
Zandile		X	X	
Jordy	X			X
Cathy		X		X
Amina	X			X
Athi	X			X
Isabella	X			X
Total: 总计：				

2. Colour in the names on the table on page 88 of Lebo's friends according to the clothes they like to wear:
根据 Lebo 的朋友所喜欢穿的衣服，将第 88 页表格中的名字涂色：

a. If they like to wear jeans and a tank top, colour their names red.
如果她们喜欢穿牛仔裤和背心，将她们的名字涂成红色。

b. If they like to wear jeans and a T-shirt, colour their names blue.
如果她们喜欢穿牛仔裤和 T 恤，将她们的名字涂成蓝色。

c. If they like to wear shorts and a T-shirt, colour their names yellow.
如果她们喜欢穿短裤和 T 恤，将她们的名字涂成黄色。

d. If they like to wear shorts and a tank top, colour their names green.
如果她们喜欢穿短裤和背心，将她们的名字涂成绿色。

3. How does colouring in the names help?
给名字涂色有什么用呢？

____

4. Which were more popular: jeans or shorts? ____________________
哪个更流行：牛仔裤或短裤？

5. Which were more popular: T-shirts or tank tops? ____________________
哪个更流行：T 恤或背心？

6. Now put the information from page 88 onto this bar graph. Use the colour combinations in Question 2 to colour your bars correctly.
现在把第 88 页的信息填在下面的条形图里。用问题 2 中的颜色组合给条形图涂色。

Popular summer clothes 流行夏装				
6				
5				
4				
3				
2				
1				
	Jeans and tank top 牛仔裤和背心	Jeans and T-shirt 牛仔裤和 T 恤	Shorts and T-shirt 短裤和 T 恤	Shorts and tank top 短裤和背心

Notes

注释

下面是帮助孩子的活动指南：

- 不管你觉得是否需要，使用真实的对象，以帮助孩子理解一个概念，例如：2个餐叉加上2个餐叉等于4个餐叉。
- 开始活动前，收集任何你需要的材料。
- 和孩子一起阅读注释。
- 孩子们能够集中精力的时间很短，所以让孩子一次做1~2页即可。

P2，P3

按倍数计数是为孩子学习乘法打下良好基础。在第二部分的活动中，孩子需要算出这些乘法的答案。可以先让他将乘法题写成重复数相加。例如：2×10=10+10=20

答案：(1)10，20，30，40，50，60；
3，6，9，12，15，18，21，24；
5，10，15，20，25，30，35，40，45，50；
2，4，6，8，10，12，14，16，18，20。
(2)40，10，6，18，15，20，9，10，15，60，25，5。

P4

为帮助孩子进行本项活动，需使用一些可以按不同方式分组的实物，例如：用10颗糖果让他分给2个人、5个人或者10个人。

答案：(1)4，10，5。(2)6，3，2。(3)12，4，6。

P5

本项活动着重于两位数的减法。提醒孩子，要用大的数减去小的数来得到答案。

答案：21，37，18，26，25，32，12，14。

P6，P7

本项活动是练习按倍数计数。孩子写出数字后，让他说出数字模式。

答案：Jody：40，45，50，55，60，65，70，75；
Mandla：24，27，30，33，36，39，42，45；
Jaco：20，30，40，50，60，70，80，90；
Lebo：60，80，100，120，140，160，180，200；
Emma：22，24，26，28，30，32，34，36；
Ravi:100，150，200，250，300，350，400，450。
Ravi募捐得钱最多。

P8

本项活动能让孩子对两位数加减法树立自信心。

答案：(1)测验1：67，64，77，99，97，68，82；
测验2：68，97，99，77，64，67，82。
(4)可以在方框中看见字母"I"。

P9

本项活动可发展孩子对数字的概念，并提供除法、减法和分数运算的练习。

答案：(1)20，20。(2)5。(3)15，3。(4)Pete得到的最多：20个金币，其次是Bill：5个金币；剩余的海盗每人3个金币。(5)$\frac{1}{4}$

P10

本项活动中，让孩子收集数据，然后放到统计图表里。一定要让孩子知道，他只能对图片中能看到的身体部位计数。

答案：4只昆虫，12条腿，6个触角，11个斑点，4只翅膀。

P11

通过用纸板制作表盘和指针，帮孩子学会报时。记住要尽可能常问他实际时钟上的时间。

答案：(2) 差15分钟到12点或者11点45分，9点，下午5点过5分，上午8:30或者是8点半。(4)14:15。

P12

本项活动中孩子要练习数字加倍或者减半。最开始可以让他用厨房里的物品加倍或者减半。一旦他做到了这点，就让他独立完成这个活动。

答案：(1)8根钓鱼竿，32个挂钩，48只幼虫，16个鸡蛋，36个小面包，28瓶饮料，24块巧克力。
(2)2根钓鱼竿，8个挂钩，12只幼虫，4个鸡蛋，9个小面包，7瓶饮料，6块巧克力。

P13

帮孩子找到容量正确的容器。例如：水罐要比碗和杯子装的水更多。最后的测量，孩子可以选择任何容器(或其他的)，但要鼓励他用小容器测量大容器的容量。

P14

本项活动中，孩子要识别和匹配图形。一定要让他能区分圆边和直线边。在(3)中，让他在给每个边框涂色之前先描述一下这个图案。

P15

开始这项活动时先让孩子数一数野餐食品。必要的话，让他把这些食品画在一张单独的纸上，然后再进行平分。提醒他，分果汁可以用分数。

答案：(1)总计：4个热狗，13个苹果，6袋薯片，2盒果汁，14块蛋糕，25个橙子。(2)每人分到：1个热狗，3个苹果，1袋薯片，$\frac{1}{2}$盒果汁，3块蛋糕，6个橙子。(3)剩下的：0个热狗，1个苹果，2袋薯条，0盒果汁，2块蛋糕，1个橙子。

P16，P17

乘法是这项活动的重点。如果孩子是用铅笔来完成这项活动，那么他可以重复做几次，每次都会有提高。

答案：赛道1：16，10，18，14，12，20，8；
赛道2：45，25，10，15，40，50，20，35，30；
赛道3：10，40，70，30，90，100，20，50，60，80；
赛道4:8，15，50，80，35，12，0，45，6。

P18

这项活动是对二维图形的识别，你可以延伸这项活动让孩子浏览一本杂志，识别并剪出所有的圆形或正方形等。

答案：(2)7个三角形，4个正方形，3个长方形，10个圆形。三角形、正方形和长方形是有直线边的。圆有圆边。正方形和长方形有相同的边数。

P19

本项活动中，孩子要用币值来解答钱数问题，包括找零钱。他还要完成表格填写。你借此机会和孩子讨论一下生活费用，尤其是养一只宠物的成本和责任。

答案：(1)狗洗发水40元，虱子粉35元，美容院打理180元，玩具100元，狗饼干300元，兽医账单200元。一年总共花费了855元。(2)3。(3)2.5元。(4)5个1元硬币。(5)狗饼干。(6)虱子粉。

P20，P21

本项活动是练习用不同倍数来计数。鼓励孩子注意每条路线的规律。还要注意按2的倍数和20的倍数计数，5的倍数和50的倍数计数，按10的倍数和100的倍数计数之间的相似之处。

答案：2，4，6，8，10，12，14，16，18，20；20，40，60，80，100，120，140，160，180，200；5，10，15，20，25，30，5，40，45，50；50，100，150，200，250，300，350，400，450，500；10，20，30，40，50，60，70，80，90，100；100，200，300，400，500，600，700，800，900，1000。

P22，P23

本项活动可提高孩子对数字的理解以及他识别奇数和偶数的能力。提醒他注意只有偶数不用分数就能等分。鼓励他在把一个数字分解成十位数和个位数时慢慢地说出来。

答案：(1)奇数/Goops：27，29，31，43，45，51，67，99；偶数/Zoops：16，24，34，38，62，72，78，88。
(2)20+7，20+9，30+1，40+3，40+5，50+1，60+7，90+9，10+6，20+4，30+4，30+8，60+2，70+2，70+8，80+8。

P24

本项活动中孩子将练习三位数排序。作为巩固练习，让孩子写出数字名称。例如：一百二十六。

答案：(2)123，126，156，179，254，290，345，374，391，421。

P25

本项活动中，孩子要自己收集朋友和家人的数据。和他一起阅读说明使他清楚如何填这个表格；每一个人要在两个方框里涂色：故事或信息，书或光盘。帮他找出哪种组合最受欢迎，哪种最不受欢迎。

P26

"重的"和"轻的"是用来描述物体的相对质量。例如：手机比羽毛重，但是比砖轻。让孩子将屋子周围的物品进行比较。用来比较的物品质量不要太相似，否则，他将难以对它们分类。

答案：轻的：茶包，铅笔，袜子，信封；重的：枕头，书，鞋，手机。

P27

本项活动可挖掘孩子对数字的概念。

答案：453=400+50+3，345=300+40+5，135=100+30+5，354 =300 +50 +4，412 =400 +10 +2，529 =500 +20 +9，434=400+30+4，783=700+80+3。

P28，P29

本项活动会教给孩子加法的技巧。鼓励他注意这种方式，用第1个数与第2个数中的一部分相加，使其成为10的整数倍，然后再把第2个数余下的差数相加。

答案：(1)19+3=19+1+2，29+5=29+1+4，9+8=9+1+7，39+5=39+1+4，49+2=49+1+1，19+9=19+1+8。(2)5，3，4，6，7。(3)16+4+5，18+2+7，37+3+2，45+5+3，78+2+4。(4)36+4+6=46，19+1+7=27，77+3+3=83，25+5+2=32，8+2+9=19。

P30

本项活动的第一部分帮助孩子认识到二维图

彡的对称性。当他完成(1)时,让他说出图形的名
≡。还要鼓励他找出家里的对称物体。帮助他区分
直边和圆边。

P31

本项活动要求孩子使用标准测量单位。确保他用
的格尺是有刻度的那边。鼓励他测量要精确到厘
米。

P32

让孩子观察这些数字,然后猜出哪一个最大。让他
猜家里的物品,给他提供额外的练习。例如:盘子
里有多少块饼干,他的盘子里有多少粒豌豆。
答案:(2)332,426,281,332,427,271。(3)第1竖
行的两把香蕉。(4)倒数第2把香蕉(427)。

P33

做所有的测量活动之前,孩子应该先估计一下。要
确保孩子测量的是长度(最长的边)而不是物品的
宽度。

P34,P35

提醒孩子用图例来理解平面图。要让他在平面图
和图例上加上箱子(11)和黑板(13)的符号。你可
以扩展这个活动,让他画出家里一间卧室的平面
图或者花园的平面图,也可以是幻想出来的房间。
答案:(1)20。(2)6。(3)10。(4)不是。(5)10。(6)
·。(7)橱柜。(8)在书桌后面/在椅子前面。(9)在他
的桌子后面和垫子前面。(10)5。(12)12。

P36

本项活动要求孩子把分数相加变成整数。如果他发
现这项活动太难,那就画一些圆,将圆分成等分,四
等分,三等分,六等分,八等分。鼓励孩子用不同的方
法使其变成整数。例如:1/2+1/4+1/4=1 和 1/2+1/2=1
答案:(1)$\frac{1}{2}+\frac{1}{2}$;$\frac{1}{4}+\frac{1}{4}+\frac{1}{2}$;$\frac{1}{4}+\frac{1}{4}+\frac{1}{4}+\frac{1}{4}$;$\frac{3}{4}+\frac{1}{4}$;$\frac{2}{4}+\frac{2}{4}$;$\frac{1}{3}+\frac{1}{3}+\frac{1}{3}$;$\frac{3}{6}+\frac{1}{2}$;$\frac{2}{3}+\frac{1}{3}$;1;$\frac{6}{6}$;$\frac{8}{8}$;$\frac{2}{4}+\frac{1}{2}$。(2)12。(3)$\frac{1}{3}$。

P37

如果你有厨房秤,让孩子用它去称一下(1)中的各
种物品重量。如果你没有,让他完成下面的加法计
算。
答案:(1)270g,490g,295g,谷物和甜豆。(2)谷物。
(3)5。(4)2。

P38,P39

本项活动提供两位数加法练习。孩子加法运算时,
一定要让他在加十位数之前先将个位数相加。
答案:(1)74,71,63,76,67,59,34,63,89。
(2)跳舞的美洲虎;懒惰的狮子;8支球队的得分都
超过35分;驯服的老虎和健谈的黑猩猩;3支球
队;跳舞的美洲虎。

P40,P41

本项活动中孩子将练习数字的加倍和减半。提醒他
数字加倍的时候是变大。确保他能理解当你加倍一
个数,是加上这个数字的本身。例如:加倍6是6+
6=12。在(2)中,孩子将这些数字减半。提醒孩子这
与加倍正好相反,当一个数字减半时,它会变小。
答案:(1)9,18,36,72,;11,22,44,88;7,14,28,56;
13,26,52,104;14,28,56,112,青蛙f跳得最高。
(2)88,44,22,11;92,46,23,11;120,60,30,15;
76,38,19,9;84,42,21,10,青蛙e跳得最低。

P42,P43

本项活动中孩子将了解钱的价值和解决钱的问
题。提醒孩子500分等于5元。
答案:(1)2元;1.5元;1元;0.25元;1元;0.5元。
(4)3.4元;2.7元;4.9元;6.1元;3.45元;2.95元。
(5)狂热焦糖。(6)香蕉果酱。

P44

本项活动用分享食物来学习除法的概念。如果他
发现有帮助,就让他做一做(2)中的算术题。
答案:(1)14个棒棒糖,22块糖果,17块软糖豆,6
块果汁软糖。(2)每个孩子7个,剩余0;每个孩子2
个,剩余0;每个孩子5块,剩余2块;每个孩子2
块,剩余3块;每个孩子10块,剩余1块;每个孩子
5块,剩余1块;每个孩子2块,剩余0;每个孩子3
块,剩余0;每个孩子3个,剩余2个;每个孩子1
个,剩余6个。

P45

提醒孩子指针式时钟是12小时制,数字时钟是24
小时制。
答案:(2)eight minutes past nine in the evening, a quarter to twelve, thirteen o' clock, half past two, a quarter past seven.(4)07:25。

P46,P47

本项活动为孩子提供了数据和图表练习。一起阅
读第46页的信息,然后让他完成这个图表。第二
个部分的活动要求孩子解读图表中的信息。
答案:(2)12;意大利香肠;橄榄;8-6=2,多2个孩
子更喜欢意大利香肠。4-2=2,多2个孩子更喜欢
鸡肉;意大利香肠,菠萝,鸡肉和蘑菇,因为这些是
最流行的口味。

P48

本项活动介绍了周长的概念。孩子需要一个小盒
子(如火柴盒)、一段长绳子、胶带和剪刀。你可以
扩展这项活动,沿着格尺放好绳子来测量盒子的
周长,然后读出测量的厘米数。鼓励孩子用测量这
个盒子的方法来测量足球场图片的尺寸。

P49

本项活动中,孩子会通过计算卡片上的加减法算
术题找到相匹配的卡片。他完成任务后,让他在成
对儿的卡片上,为每对儿涂上不同的颜色。
答案:(1)360,369,291,621,360,369,621,291。
(3)291。

P50,P51

本项活动提供了更多的分数练习。必要时,让孩子
用纸板代替披萨,用豌豆代替坚果。
答案:(2)$\frac{1}{2}$或者$\frac{4}{8}$,$\frac{3}{4}$或者$\frac{6}{8}$,$\frac{8}{8}$,$\frac{3}{8}$,$\frac{6}{8}$或者$\frac{3}{4}$,$\frac{1}{4}$或者$\frac{2}{8}$。(3)Jaco。(4)Emma。(5,6)Jody:$\frac{1}{2}$,6;Mandla:$\frac{3}{4}$,9;Jaco:1,12;Ravi:$\frac{1}{3}$,4;Lebo:$\frac{2}{3}$,8;Emma:$\frac{1}{4}$,3。(7)$\frac{1}{4}$,$\frac{1}{2}$,$\frac{3}{4}$,1。

P52

本项活动的重点是三位数与两位数减法。必要时,
可以用一张纸来算出答案。
答案:(1)647-25=622,198-76=122,473-62=
411,281-51=230,356-33=323,529-16=513。
(2)袋子里最多剩622粒花生。
(3)袋子里最少剩122粒花生。

P53

本项活动中孩子要识别出不同图片中的二维图
形。
答案:鸡蛋:在这个鸡蛋上的图案有椭圆形、三角
形、正方形;鸟:头是圆形,身体是椭圆形,鸟嘴、脚
和羽毛是三角形;机器人:天线是由圆组成,头是
一个六边形,眼睛和脖子是正方形,脚、嘴和按钮
是长方形;冰淇淋:蛋卷是一个三角形,樱桃是圆
形;电脑:屏幕和键盘是长方形,在屏幕上有六边
形、长方形、正方形和圆形。

P54,P55

这项测试是让孩子进行时间转换练习。他可能会
发现,用钟表和日历有助于求出答案。
答案:(1)$\frac{3}{2}$,24小时+6小时=30小时。(2)3个月,
30+31+31=92天。(3)60+60+45=165分钟,$\frac{11}{4}$小
时或者2小时45分钟。(4)30分钟,半小时。(5)5
小时,25小时。(6)28+31+11=70天。

P56

需要花些时间提醒孩子分数可以写成不同的方
式。例如,$\frac{1}{2}=\frac{2}{4}$。可以通过切水果或剪纸板来演
示一下。
答案:(1)8。(2)$\frac{4}{8}$或$\frac{2}{4}$或$\frac{1}{2}$。(3)$\frac{2}{8}$或$\frac{1}{4}$。(4)$\frac{3}{8}$(5)$\frac{2}{8}$或$\frac{1}{4}$。(6)$\frac{6}{8}$或者$\frac{3}{4}$。(7)$\frac{4}{8}$或者$\frac{1}{2}$或者$\frac{2}{4}$。(8)全部或者1。

P57

这项活动教给孩子的概念是,乘法中的两个数可
以调换位置,而不影响答案。
答案:(1)3,×,6,=,3,×,×,10。

P58,P59

活动的第一部分是要求孩子将重复的数字相加。
第二部分中,要从大到小将三位数排序。
答案:(1)有多种正确的组合。下面是每个飞镖的
两个例子:100+100+100+50+20+10/50+50+50+50+
50+50+50+20+10;100+100+50+10/100+100+20+
20+20;100+100+100+100+100+50+20+20/100+
100+100+100+100+20+20+20+20+5+5;100+100+
100+20+10+5/100+100+50+50+20+10+5。(2)
410,195,520,340。(3)672,651,632,543,456,452,
446,432,430,412,352,345,291,267,189,165,145,
139。(4)672。(5)139。

P60,P61

本项活动给孩子提供了一个设计平面图的机会。
在(2)中,提醒他在给平面图上添加物品时,图例
中的每个物品至少都要用1个。在他画平面图之
前,让他花些时间注意一下他的卧室里的物品在
什么位置。(3)在孩子完成这个平面图之后,鼓励
孩子检查一下图例是否完整。提醒他,他可以使用
任何符号,只要平面图和图例上的符号相同即可。

P62,P63

本项活动提供了进一步加倍和减半的练习。一起
阅读说明,确保孩子能理解每次要求解决的问题。
答案:(1)426cm,382cm,306cm,268cm。(2)852cm,
764cm,612cm,536cm。(3)213cm,191cm,153cm,134cm。
(4)新房子:692升,968升,300升,528升;Terry的

房子：173 升，242 升，75 升，132 升。

P64

本项活动中，要求孩子用除法算出每个队多次比赛的得分。如果需要，解释一下每个队得分应除以比赛的得分类型 2、4 或 6。最后一部分活动要求他将各队按得分最多到最少排序。

答案：(1)6,7,10,12。(2)黑猩猩，松鼠，浣熊，青蛙。

P65

提醒孩子什么是周长，我们怎样用绳子去测量它。(你可以参考第 48 页。)然后让他用一根绳子仔细测量每个图形的轮廓。让他将绳子贴在格尺的厘米标记上。确保绳子超始端在格尺的 0cm 处。你可以扩展这项活动，让孩子再用格尺测量一下这些图形。所得答案应该和之前用绳子测得的相同。

答案：(2)12cm，11cm，20cm，20cm。(3)最后两个图形的周长最长(20cm)。

P66，P67

本项活动可以引起孩子注意我们周围环境中各种物体的形状。鼓励他在日常环境中注意其他物品的形状，并说出其名称。

答案：(1)锥体，球体，长方形。(2)圆形，长方形，三角形，正方形。(3)长方形，正方形。(4)菱形，球体，圆形，立方体，五边形。(5)长方形，三角形，圆形，正方形。

P68

本项活动提供了乘法练习。提醒孩子乘法可以解释为几组数字，例如：$3\times4=3$ 组 4。深一步练习，鼓励他做出他自己的乘法捉对儿卡片。

答案：(1)$3\times4=6\times2$，$5\times4=2\times10$，$12\times2=4\times6$，$3\times3=9\times1$，$4\times4=8\times2$。(3)3×7。

P69

本项活动让孩子练习十位和百位的加减法。让他将炮弹与正确的大炮连线。

答案：(1)637=737−100/627+10/537+100/647−10；425=325+100/225+200/525−100；863=853+10/763+100/873−10。(2)大炮 637。

P70，P71

提示孩子计数时可以从任何一个数开始，按照规律数。可以扩展这个活动，让他从不同的数字开始，按不同的倍数计数。例如：从 12 开始按 3 的倍数计数；现在从相同的数字开始，按 5 的倍数计数。

答案：(1)第一个海盗：3,6,9,12,15,18,21,24, 27, 30；第二个海盗：4,8,12,16,20,24,28,32,36,40；第三个海盗：100,200,300,400,500,600,700,800,900,1000；第四个海盗：25,50,75,100,125,150,175, 200,225,250。(2)鹦鹉 69,66,63,60,57,54,51。(3)1002,2002,3002,4002,5002,6002,7002,8002,9002；1061,2061,3061,4061,5061,6061,7061,8061, 9061。

P72

本项活动提供了加法练习。你可以在乘车中玩汽车号牌游戏。让孩子对周围汽车的号牌上的数字做加减乘法运算。

答案：(1)$246+3=249$，$167+1=168$，$752+6=758$，$641+2=643$，$340+7=347$，$931+2=933$，$872+4=876$，$943+1=944$。(2)944。(3)168。

P73

本项活动提供了减法练习。鼓励孩子注意这个图案。在(2)中，建议他在安全的船旁边打个“√”号，在要下沉的船旁边打个“×”。

答案：(1)463，442，221，584，473，642，261，219。(2)2 只船会下沉；6 只船是安全的。

P74

本项活动中，孩子将学会带有分数的时间。

答案：(1)15 分钟，30 分钟，45 分钟，60 分钟。(2)$\frac{1}{2}$，$\frac{3}{4}$，1，$\frac{1}{4}$。

P75

本项活动建立在第 74 页的基础上。如果需要让孩子参照一下第 74 页。提示他分数时间不一定开始于 12 点。

答案：(1)1 小时的$\frac{3}{4}$。(2)15 分钟。(3)60 分钟或者 1 小时。(4)30 分钟或者$\frac{1}{2}$小时。

P76，P77

本项活动可扩展孩子的数字概念，并提供加法练习，让孩子用三种方法拆分金字塔顶端的数字。确保他能理解金字塔每一行的数字相加起来应等于顶端的数字。

P78

如果你有一个可用的漏斗，让孩子用它来浇注会更容易，更精确。确保他把每一个杯子都正确装满，以确保更精确的测量。当孩子完成了活动，花一些时间讨论一下他的估计值，及其近似程度。

答案：(3)5，10，6，4，8。

P79

本项活动是练习除法计算，但每个答案都有余数。你要提醒孩子，蜘蛛有 8 条腿，蜜蜂有 6 条腿。

答案：(1)8。(2)6。(3)4。(4)2。(5)4，1。(6)6，2
(7)5，1。(8)14，1。

P80，P81

本项活动提供了重要的乘法口决练习，在(1)中一定要让孩子知道，他先要算出第一个计算题的答案，才能知道缺少的数字是什么。

答案：(1)$20\times5=10\times10$，$6\times2=3\times4$，$6\times10=10\times6$，12
3=6×6，$4\times9=3\times12$，$2\times2=1\times4$，$5\times4=10\times2$，$2\times12=6$
4，$10\times3=5\times6$，$2\times9=3\times6$。(2)$7\times4=28$，$12\times3=36$，6×6
36，$3\times2=6$，$1\times10=10$，$2\times9=18$，$4\times6=24$，$5\times9=45$，8
8=64，$9\times6=54$。

P82，P83

本项活动中孩子要先找出信息(数据)，然后将它记录在表格里。要将这些信息转换成一个条形图确保他明白他需要在表格中将相应的方框涂上颜色，以反映出在第 1 个表格里 6 个人各自的选择。

P84，P85

本项活动提出一些问题，让孩子练习用除法去解答。

答案：(1)16,6,12,8,10,30,22。(2)8,3,6,4,5,15
11。(3)24,9,18,12,15,45,33。

P86

本项活动是给孩子一个测量面积的实际体验。你可以扩展这项活动，测量一下窗户或墙的面积。你用的纸张可能不完全适合。如果报纸的尺寸太大试着用半张报纸进行测量。

P87

本活动以前一页为基础。帮孩子按提供的尺寸制作马赛克纸板砖和地垫底部。你需要测量、剪裁并准确粘贴。

答案：(5)10 块。

P88，P89

本项活动要求孩子解释表中信息，并将这些数据转换到条形图上。鼓励孩子在完成图表时所用的颜色与服装的颜色相同。

答案：(1)牛仔裤：8，短裤：4，T 恤：3，背心：9。(3)更容易看出不同衣服的选择。(4)牛仔裤。(5)背心。(6)牛仔裤和背心：涂 6 个方框；牛仔裤和 T 恤：涂 2 个方框；短裤和 T 恤：涂 1 个方框；短裤和背心：涂 3 个方框。

Certificate

证书

已通过了《奔跑吧,数学》4 级闯关!

Smart~Kids

Make the smart choice for a brilliant future!

Smart-Kids Mathematics
英汉对照
奔跑吧数学
1级
Grade 1
探索奇妙的数学世界
〔英〕 罗宾·布莱斯
安妮·海瑟薇 主编
许译丹 译
Robyn Brice, Anne Hathway
Pearson

Smart-Kids Mathematics
英汉对照
奔跑吧数学
2级
Grade 2
探索奇妙的数学世界
〔英〕 卡罗尔·艾瑞
米歇尔·福尔 主编
赵媛媛 译
Carol Every, Michelle Faure
Pearson

Smart-Kids Mathematics
英汉对照
奔跑吧数学
3级
Grade 3
探索奇妙的数学世界
〔英〕 卡罗尔·艾瑞
吉恩·彼得斯 主编
赵媛媛 译
Carol Every, Gené Peters
Pearson

Smart-Kids Mathematics
英汉对照
奔跑吧数学
4级
Grade 4
探索奇妙的数学世界
〔英〕 吉恩·彼得斯
卡罗尔·艾瑞 主编
许译丹 译
Gené Peters, Carol Every
Pearson

Stickers

P 10

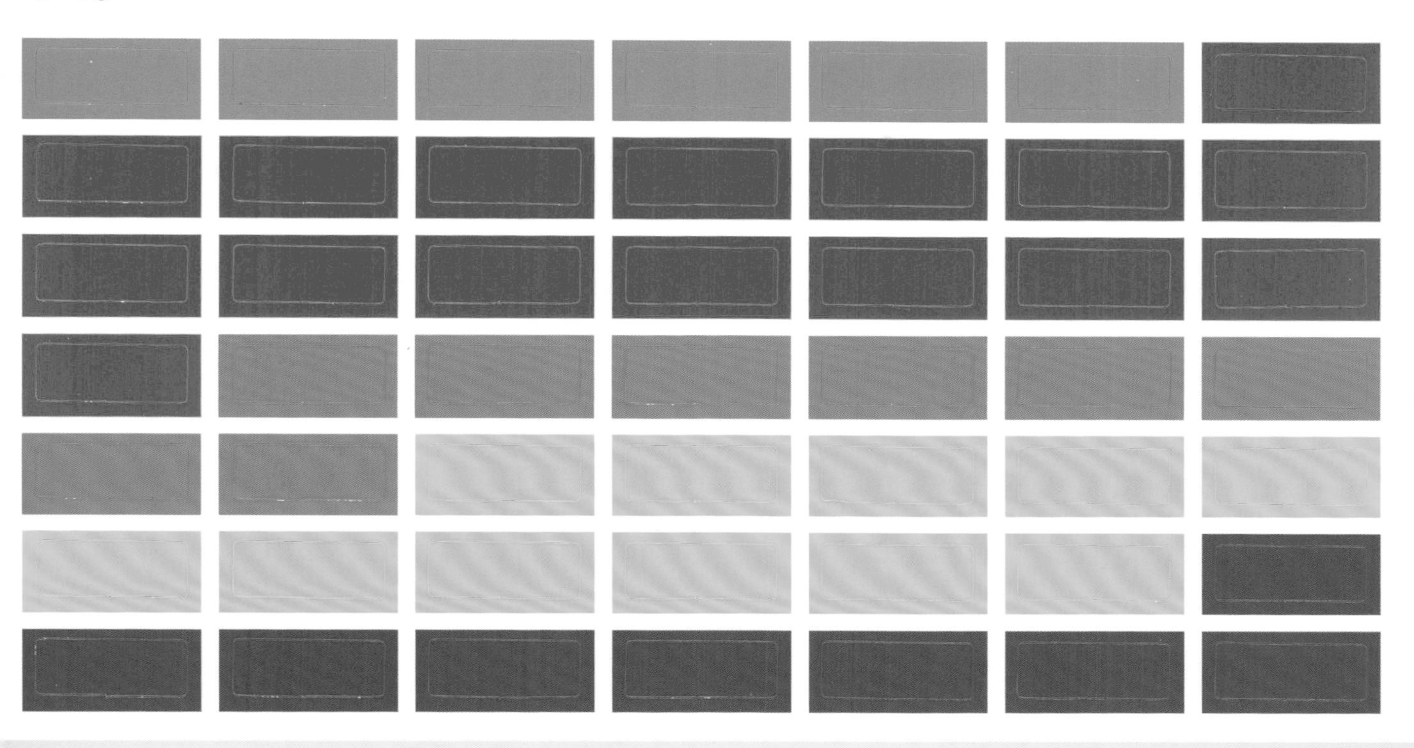

P 14

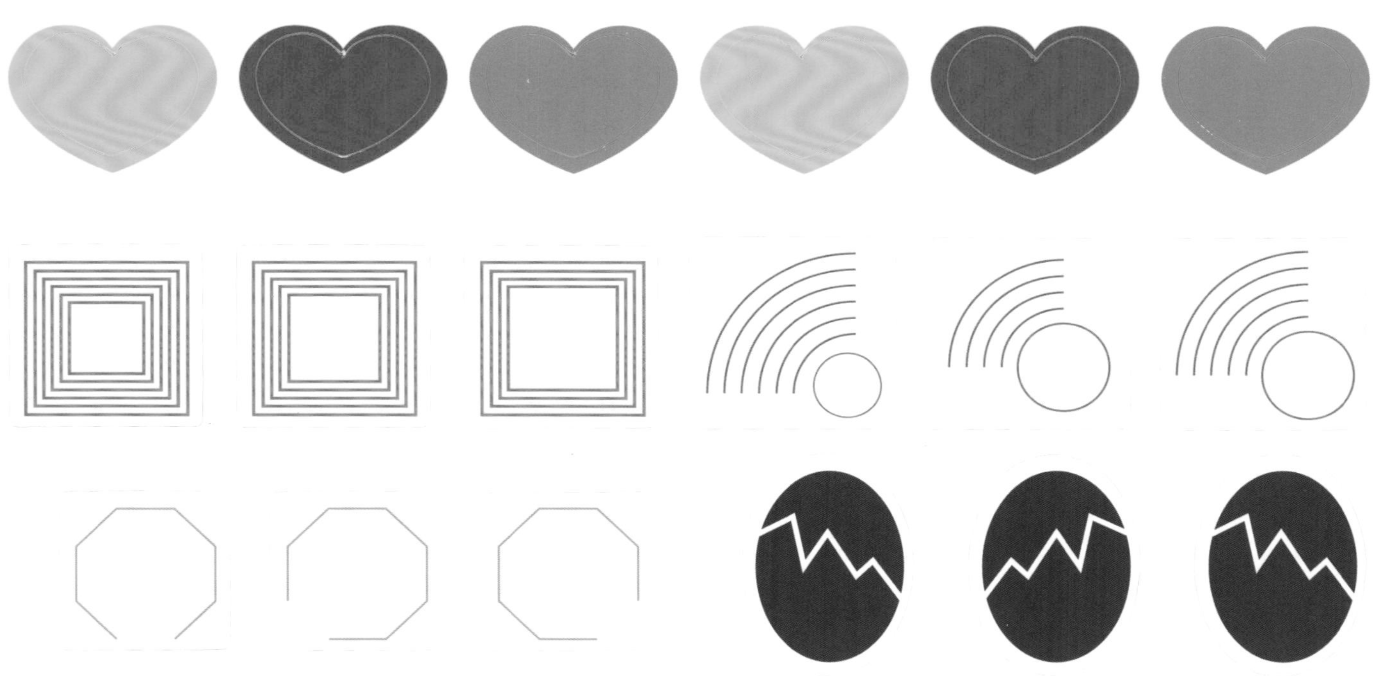

P 24

Stickers

P 47

 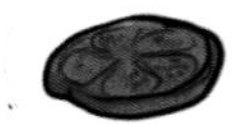  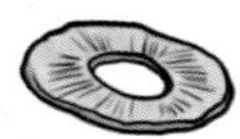  

P 50

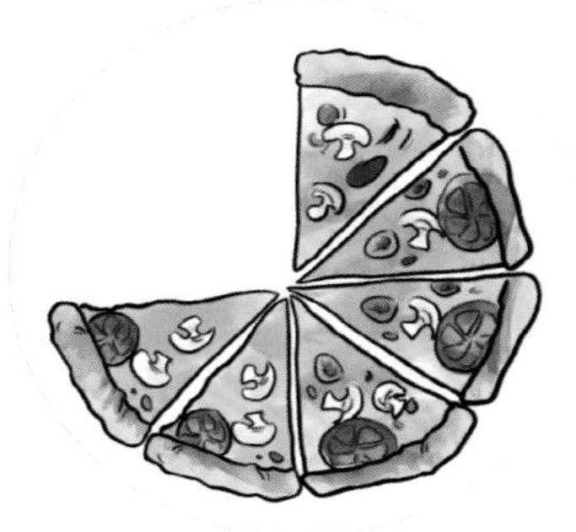  

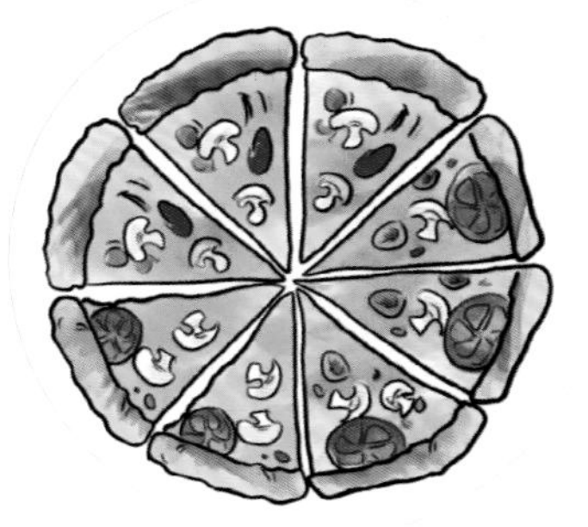 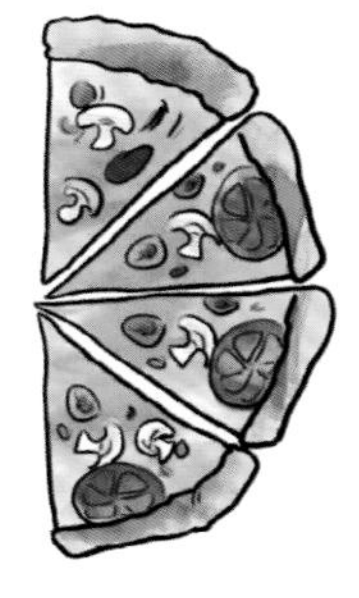